DIVISIONS of the BRITISH ARMY

1939 - 45

Compiled & Published by

Mr. M.A. Bellis
10 White Hart Lane
Wistaston, Crewe,
Cheshire, CW2 8EX

53rd (Welsh) Division

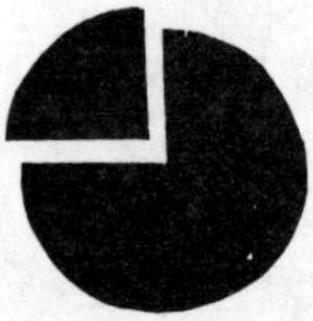

4th Division

50th (Northumbrian) Division

44th (Home Counties) Division

6th and 70th Divisions

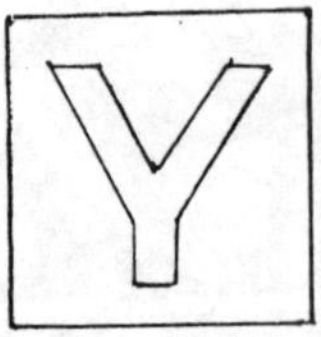

5th Division

3rd Division

DIVISIONS OF THE BRITISH ARMY 1939 - 45

INTRODUCTION

A continuing interest in British Military Vehicles during the period 1939-45, embracing both full-size reconstructions and a variety of scale models, has motivated the publication of this Booklet. It comprises a collation of data from a number of the definitive publications on this and related subjects, including selected extracts from the Army Lists, Orders of Battle of WWII (Joslen H.F.), Formation Badges of WWII (Cole H.), Lineage Book of the British Army (Frederick J.B.), Military Markings of WWII (Wise T.), and British Military Markings of WWII (Hodges P.).

Divisions of the British Army 1939-45 concentrates on the Armoured, Infantry and Artillery contingents of the Division, taking the smallest recorded unit (except in a few instances), as the Regiment or Battalion. The Booklet does not cover periods when these units served with independent Brigades or with the higher formations (Corps or Army Troops).

In providing information which will enable the correct colour schemes and markings to be applied, it has been necessary to be able to determine unit seniorities when brigaded together, this necessitating some research into the History of the British Army. This data has been compiled into supplements 1,2 & 3, which can be cross-referenced with arm-of-service markings in tables 1 & 2. Additionally, figures 1 to 6 give diagramatic representations of simplified command structures whilst the Index enables individual Regiments/Battalions to be traced back to parent Divisions.

Malcolm A. Bellis

DIVISIONS OF THE BRITISH ARMY 1939 - 45

Relationship to Corps and Army

The Division was so placed in the Command structure that it was often the means of converting strategic planning into terms of practical military action. The Division was immediately subordinate to the Corps, which may have contained between one and five divisions, depending on the nature of the task and unit availability. It was not unusual for a Corps to contain Divisions of different nationalities. In turn, a Corps was subordinate to the Army (used as a formation title, not a generic term). Again, there seems to have been no definitive constitution for an Army, and the number of subordinate units again varied according to cicumstances. This variation in Corps and Army constitution is shown in figure 1, which illustrates three Armies considered to be typical.

Types of Divisions employed

The two main types of Division were the Armoured Division and the Infantry Division. Both continued to evolve between 1939 and 1945, with the Armoured Division probably being subjected to the greater number of changes. Both these types will be examined in more detail. It is interesting to note that both the Armoured and Infantry Division organisations promoted the formation of 'Specialist' Divisions, the Armoured Assault Division eminating from the Armoured Division and the Airborne Divisions from the Infantry Division. The Cavalry Division, of which only one remained by the outbreak of war, was still horsed cavalry, not having completed its intended mechanization programme. Despite its lack of mechanisation, the Division still had approx. 1800 motor vehicles in addition to its 6000 horses and 11,000 personnel. Following the occupation of Vichy French Syria, the Division was to be disbanded, having commanded the last British horsed-cavalry engaged on active service.

The Armoured Divisions

In 1939 only two Armoured Divisions existed, the 1st serving with the B.E.F. in France & Belgium and the 2nd remaining in the UK. In Egypt the Mobile Division, with an even more acute shortage of equipment than its European counterparts, continued to evolve to its later designation of 7th Armoured Division, redesignated in February 1940.

FIGURE 1

EXAMPLES OF ARMY * CORPS * DIVISION RELATIONSHIPS

ARMY	CORPS	DIVISIONS ETC.
British 8th Army El Alamein October 1942	X Corps	1st Armoured Division 8th Armoured Division 10th Armoured Division
	XXX Corps	4th Indian Division 51st Infantry Division 2nd New Zealand Division 1st South African Division 23rd Armoured Brigade Group
	XIII Corps	7th Armoured Division 44th Infantry Division 50th Infantry Division 1st & 2nd Free French Groups
British 10th Army Persia and Iraq March 1943	III Corps	6th Indian Infantry Division
	XXI Indn. Corps	8th Indian Infantry Division 56th Infantry Division
British 1st Army Tunisia May 1943	V Corps	78th Infantry Division 1st Infantry Division 4th Infantry Division
	IX Corps	6th Armoured Division 46th Infantry Division 1st Armoured Division 7th Armoured Division 4th Indian Division 201st Guards Brigade
	US II Corps	US 1st Armoured Division US 1st Infantry Division US 9th Infantry Division US 34th Infantry Division
	French XIX Corps	Div. du Constantin

Particular note should be paid to both the number of subordinate units and the very different 'mixes' of Division types, particularly in relation to Armoured and Infantry units. In addition to the units detailed, Armies and Corps had units reporting directly to their headquarters. These are referred to as 'Army Troops''and 'Corps Troops' respectively.

FIGURE 2 — MAJOR CHANGES IN ARMOURED DIVISION ORGANISATION 1939-45

1939

HQ Armd.Div.

- HQ Armd.Bde
  - 3xArmd.Rgts
- HQ Armd.Bde
  - 3xArmd.Rgts
- HQ Support Group
  - 2x Mot.Bns
  - RHA Rgt
  - Atk/LAA Rgt
  - Engineers
- Others
  - R.A.S.C
  - Medical
  - Signals
  - Provost

1940

HQ Armd.Div.

- HQ Armd.Bde
  - 3xArmd.Rgts
  - 1x Mot.Bn
- HQ Armd.Bde
  - 3xArmd.Rgts
  - 1x Mot.Bn
- HQ Support Group
  - RHA Rgt
  - Atk Rgt
  - LAA Rgt
  - 1x Inf.Bn
- Others
  - Engineers
  - R.A.S.C.
  - Medical
  - Signals
  - Provost

1942 (Middle East)

HQ Armd.Div.

- HQ Armd.Bde Group
  - 3xArmd.Rgts
  - 1x Motor Bn
  - 1x RHA Rgt (incl.Atk.Bty)
  - LAA.Bty
  - Engineers
- HQ Motor Bde Group
  - 3 Motor Bns
  - RHA/Field Rgt (incl.Atk.Bty)
  - LAA.Bty
  - Engineers
- Others
  - HQ Div.RA
  - HQ Div.RE
  - R.A.S.C
  - Medical
  - Signals
  - Provost

Late 1942 onward

HQ Armd.Div.

- HQ Armd.Bde
  - 3xArmd.Rgts
  - 1x Motor Bn
- HQ Inf.Bde
  - 3x Inf.Bns
- HQ Div.RA
  - 2x RHA/Fld Rgts
  - Atk. Rgt
  - LAA. Rgt
- A/Car Rgt
- Others
  - Engineers
  - R.A.S.C.
  - Medical
  - Signals
  - Provost

Further changes to 1945

(i) A/Car Rgts. to Corps Troops, replaced by Armd. Recce. Rgts. 1943

(ii) Machine Gun Coy. added to Divisional Troops, early 1944

The Armoured Divisions (Contd..)

Following the Dunkirk evacuation, a programme to increase the number of Armoured Divisions was put into effect, resulting in three new Divisions in the UK by the end of 1940 (plus the redesignation of the Mobile Division), a further three in the UK and one in the Middle East by the end of 1941 and one new Division in the UK during 1942. Between 1939 and 1944, the total personnel of the Armoured Division grew from 9,442 to 14,964, and, although the total number of tanks reduced from about 350 in 1939/40 to about 300 in 1944/45, the latter included only a small number of Light Tanks, which had comprised a significant proportion of the vehicles of 1940. Motor transport in the categories of trucks and lorries increased from about 1320 vehicles to 1930 vehicles, the 8cwt, 1ton & 30cwt classes being almost entirely replaced by the 15cwt & 3ton classes. Carriers increased from about 90 to 260 per Division.

Organisation of the Armoured Division

Figure 2 demonstrates four important stages in the evolution of the Armoured Division. These changes were primarily motivated by the need to provide an effective balance between armour and its infantry support, and to ensure that the control of the deployment of supporting artillery was at its most efficient. The effectiveness of efficient inter-arm co-operation had been well demonstrated by the German Army during the Blitzkreig period. However, it was not until late 1942 that Divisional organisation was considered to be effective.

The Armoured Assault Division

Following the disasterous Dieppe raid of August 1942, it became obvious that considerable thought and vehicle development was necessary if any assault on 'Fortress Europe' was to be successful. To this end, the last of the Armoured Divisions to be raised, the 79th, was converted to an Armoured Assault Division, responsible for the development, testing and subsequent crew training of vehicles designed to overcome anticipated obstacles (both natural and man-made)encountered during an assault. The new Division was commanded by Maj.Gen. Sir Percy Hobart, and the variety of specialist vehicles derived became known as 'Hobart's Funnies'. These included Duplex Drive 'swimming' tanks, special R.E. vehicles capable of laying bridges or firing 'flying dustbin'demolition charges, Crab Flail

FIGURE 3

INFANTRY & AIRBORNE DIVISIONS

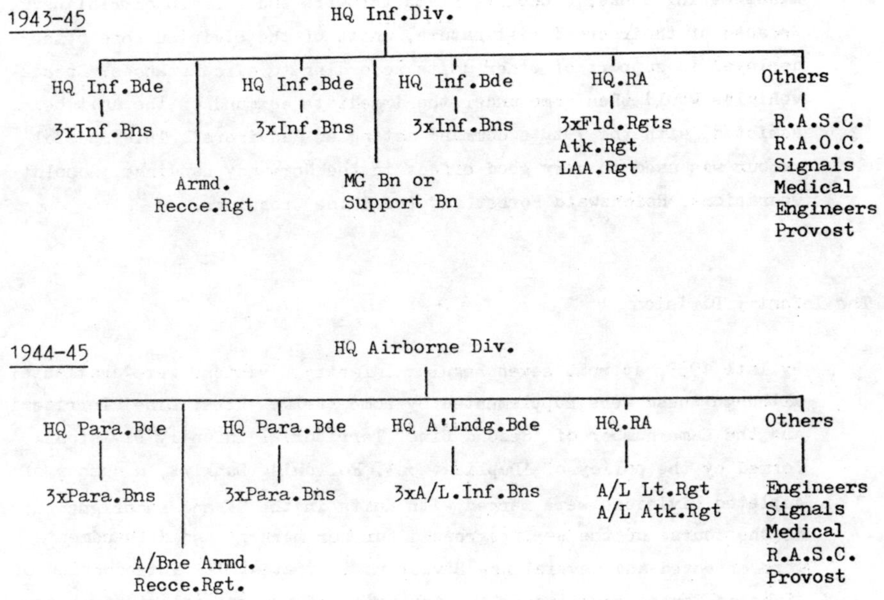

FIGURE 4

RANKS & STAR PLATES (Star plates used from 1944)

UNIT	COMMAND RANK	STARS
Army	General	Four
Corps	Lieutenant-General	Three
Division	Major-General	Two
Brigade	Brigadier	One
Regiment or Battalion	Lieutenant-Colonel	-

Vehicle plates carried yellow stars on red background

The Armoured Assault Division (Contd..)

minesweeping tanks, Crocodile flame throwers and Buffalo amphibians. Because of their specialist nature, units of the Division were often deployed in support of other units according to circumstances. Special vehicles would then come under the immediate command of the unit being assisted, with the 79th's command acting as 'advisors'. This Special armour was used to very good effect in the Normandy Landings, Scheldt Operations, Reichswald Forest and the Rhine Crossings.

The Infantry Division

By late 1939, approx. seven Regular Infantry Divisions were available, although these were supplemented by some twelve 'First Line' Territorial and the same number of 'Second Line' Territorial Infantry Divisions, formed by the policy of 'Duplication'. Following Dunkirk, a number of depleted Divisions were merged with units in the UK and redesignated. As the course of the war progressed further mergers and disbandments were effected and several new Divisions initiated. The introduction of Airborne Forces resulted in the formation of two specialist Divisions, 1st and 6th Airborne, commanding both parachute and glider-borne troops. In 1939 an Infantry Division had about 13.800 personnel of all ranks, which rose to about 18,300 by 1944, and some 12,100 for an Airborne Division. As in the Armoured Division, 8cwt, 1ton and 30cwt trucks and lorries almost disappeared, being replaced by the 15cwt and 3ton classes. Another noticeable point was the increase in carriers from 140 in 1939 to almost 600 in 1944.

Organisation of the Infantry Division

The Infantry Division changed little in its organisation during the period 1939-45, although two particular points of interest are the evolution of Reconnaissance Battalions to replace the Divisional Cavalry transferred to form 6th Armoured Division- 1940, (these battalions were redesignated regiments in 1942 and were incorporated into the RAC in Jan. 1944), and the trial of the 'Mixed Division' in 1942 where the third Infantry Brigade was replaced by a Tank Brigade in order to attempt to improve the interaction of armour and infantry (the five UK-based Mixed Divisions involved returned to Infantry Divisions in 1943). Figure 3 shows typical organisations for Infantry and Airborne Divisions.

FIGURE 5

DIVISIONAL ARTILLERY REGIMENTS

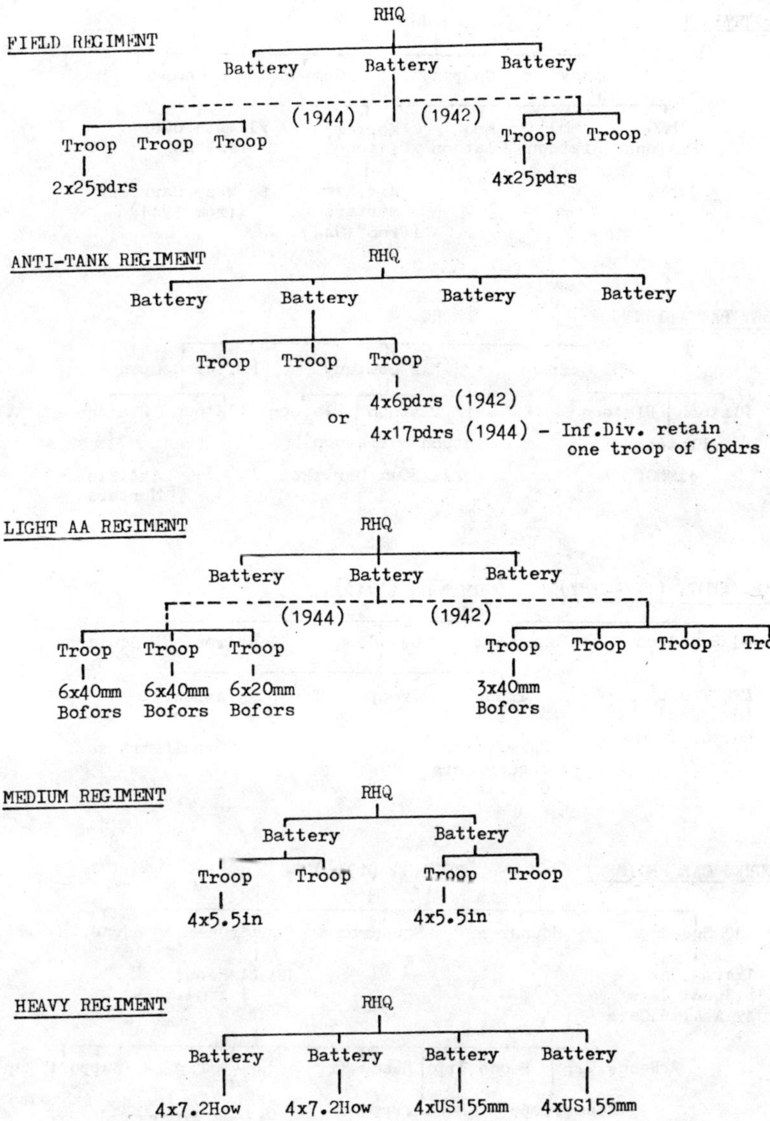

FIGURE 6 DIVISIONAL SUPPORT REGIMENTS

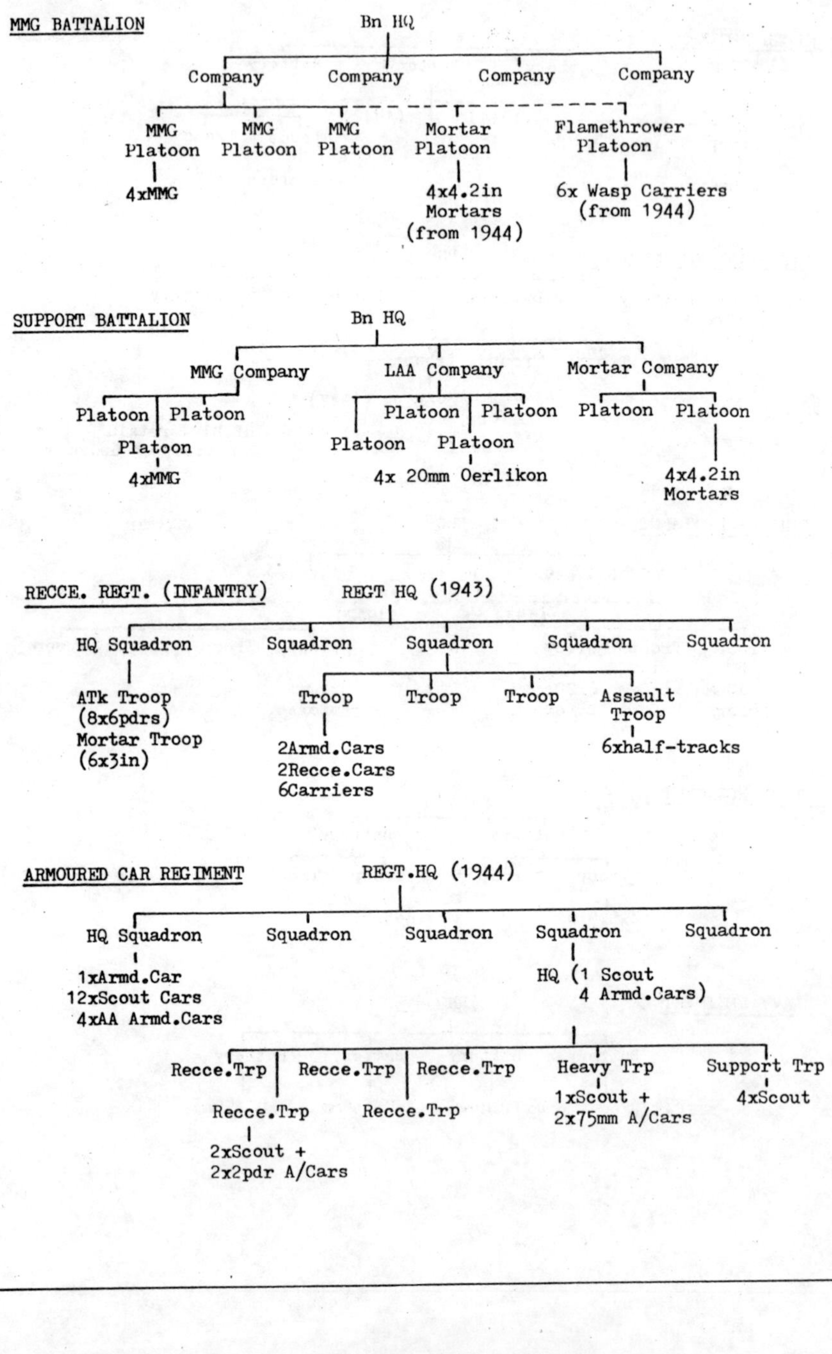

DIVISIONS - ORDER OF BATTLE

Including (in following order)

* 1st Cavalry Division
* Armoured Divisions 1-2, 6-11, 42, 79, Guards
* Infantry Divisions 1, 1London, 2, 2London, 3-6, 8, 9, 12, 15, 18, 23, 36, 38, 42-56, 59, 61, 66, 70, 76-78, 80, 1 & 6 A'Borne, Guards

The following County Divisions, employed in the UK between February and December 1941 on Coastal Defence Duties have not been included :

Devon & Cornwall (see 77th Inf.Div.), Durham & North Riding, Dorset, Essex, Hampshire, Lincolnshire, Norfolk (see 76th Inf.Div.), Northumberland, West Sussex, Yorkshire.

The following Divisions are considered to have been non-operational, for reasons stated :-

7th Infantry Division	- regular Division in Palestine, internal security duties, redesignated 6th Inf.Div. November 1939
7th Division (Cyprus)	- raised for deception purposes (6.41-6.43)
8th Division (Syria)	- Internal Security and Administration duty under command of 9th Army (6.42-10.43)
12th Division (SDF)	- Comprised entirely of units of Sudan Defence Force, no Brigades. Employed on deception & internal security duties.
40th Infantry Divn.	- Formed in Sicily for deception purposes, (11.43-6.44)
57th Infantry Divn.	- Formed in N.Africa for deception purposes, (11.43-7.44)

The following Divisions were not formed :-

Armoured : 3, 4, 12-41, 43-78

Infantry : 10, 11, 13, 14, 16, 17, 19-22, 24-35, 37, 39, 41, 58, 60, 62-65, 67-69, 71-75, 79, 2-5 Airborne

1st CAVALRY DIVISION - Formed in UK Oct.1939 as Horsed-Cavalry Division, converted and redesignated 10th Armoured Division in Syria, Aug.1941

INSIGNIA None recorded

THEATRES UK (10.39-1.40), Palestine (1.40-5.41), Transjordan (5.41), Iraq (5.41-6.41), Syria (6.41-7.41)

DIVISIONAL TROOPS 104RHA, 106RHA, 107RHA (11.39-8.40)

BRIGADES 4th CAVALRY BRIGADE (10.39-7.41)
Household Cavalry Regiment (11.39-7.41), North Somerset Yeomanry (11.39-3.41), Royal Wiltshire Yeomanry (12.39-10.40) (1.41-7.41), Warwickshire Yeomanry (3.41-7.41)

5th CAVALRY BRIGADE (10.39-7.41)
Yorkshire Hussars (10.39-3.41), Yorkshire Dragoons (10.39-7.41), Notts.(Sherwood Rangers) Yeomanry (10.39-2.41), North Somerset Yeomanry, Cheshire Yeomanry (3.41-7.41)

6th CAVALRY BRIGADE (10.39-7.41)
Warwickshire Yeomanry (10.39-3.41), Cheshire Yeomanry (10.39-3.41), Staffordshire Yeomanry (10.39-4.41)(6.41-7.41), Royal Wiltshire Yeomanry (10.40-1.41), Yorkshire Hussars, Greys (3.41-7.41)

1st ARMOURED DIVISION - Regular Division 1939, disbanded in Italy, Jan.1945

INSIGNIA — Standing white rhino, green background (later black oval), rhino depicted charging from early 1942

THEATRES — UK (9.39-5.40), France (5.40-6.40), UK (6.40-8.41), N.Africa (11.41-5.44), Italy (5.44-1.45)

DIVISIONAL TROOPS — 12th Lancers (11.40-5.42)(9.42-4.44), Royals (5.42-9.42), 4th Hussars (5.44-9.44), 2RHA & 11RHA (8.42-9.44), 4RHA (9.42-10.43), 60Atk.RA (4.44-9.44), 76Atk.RA (9.42-3.44) 42LAA.RA (9.42-10.44)

BRIGADES

2nd LIGHT ARMOURED BRIGADE (3.39-4.40) redesignated as :-
2nd ARMOURED BRIGADE (5.40-10.44)

Bays, 9th Lancers, 10th Hussars (9.39-10.44), Yorkshire Dragoons (9.42-11.42), 1Bn Kings Royal Rifle Corps (8.42 & 6.43-10.43), 9Bn Kings Own Yorkshire Light Inf. (12.42-6.43)

1st HEAVY ARMOURED BRIGADE (9.39-4.40) redesignated as :-
3rd ARMOURED BRIGADE (5.40-10.40)

2RTR (9.39-4.40)(5.40-7.40), 3RTR (9.39-6.40)(10.40), 5RTR (9.39-6.40)

22nd ARMOURED BRIGADE (10.40-11.41)

2nd Royal Glos.Hussars, 3rd & 4th County of London Yeomanries (10.40-11.41), 2Bn Kings Royal Rifle Corps (12.40-8.41)

200 GUARDS BRIGADE (3.42-5.42) redesignated as :-
201 GUARDS MOTOR BRIGADE (5.42-7.42) - Captured at Tobruk

3Bn Coldstream Guards, 2Bn Scots Guards, 9Bn Rifle Brigade (3.42-7.42)

7th MOTOR BRIGADE (10.42-7.43) redesignated as:-
18th INFANTRY BRIGADE (7.43-10.44)

1Bn Buffs, 14Bn Foresters, 9Bn Kings Own Yorkshire Light Infantry (10.42-10.44)

SUPPORT — 1st SUPPORT GROUP (9.39-2.42)

1Bn Rifle Brigade, 2Bn Kings Royal Rifle Corps (9.39-4.40), 1RHA, 2RHA (9.39-10.39), 60Atk.RA (12.39-2.40), 11RHA (8.40-1.42), 101LAA/Atk.RA (2.40-10.40), 61LAA.RA, 76Atk.RA (11.40-1.42), 82LAA.RA, 102Atk.RA (1.42-2.42), 9Bn Foresters (1.41-8.41), 2Bn Kings Royal Rifle Corps (8.41-2.42)

2nd ARMOURED DIVISION - Formed in UK Jan.1940, disbanded in N.Africa May 1941 following heavy losses in action.

INSIGNIA — Plumed knight's helmet in white on a red background

THEATRES — UK (1.40-10.40), N.Africa (1.41-5.41)

DIVISIONAL TROOPS — Kings Dragoon Guards (1.41-5.41)

BRIGADES — 1st LIGHT ARMOURED BRIGADE (1.40-4.40) redesignated as :-
1st ARMOURED BRIGADE (4.40-2.41)

Kings Dragoon Guards (1.40-12.40), 4th Hussars (1.40-2.41), 3rd Hussars (1.40-8.40), 3RTR (9.40-10.40)(2.41)

22nd HEAVY ARMOURED BRIGADE (1.40-4.40) redesignated as :-
22nd ARMOURED BRIGADE (4.40-10.40)

2nd Royal Glos.Hussars, 3rd & 4th County of London Yeomanries (1.40-10.40)

3rd ARMOURED BRIGADE (10.40-1.41)(3.41-4.41)

3RTR (10.40-1.41), 5RTR (10.40-1.41)(3.41-4.41), 6RTR (2.41-4.41), 3rd Hussars (2.41-4.41), 1Bn Tower Hamlets Rifles (2.41-3.41)

SUPPORT — 2nd SUPPORT GROUP (3.40-4.41)

1Bn Rangers (3.40-2.41), 1Bn Tower Hamlets Rifles (3.40-2.41), 102LAA/Atk.RA (3.40-2.41), 12RHA (4.40-8.40), 2RHA (7.40-2.41), 102Atk.RA (2.41), 15LAA.RA (2.41), 104RHA (3.41-4.41)

<u>6th ARMOURED DIVISION</u> - Formed in UK Oct.1940, still operational Sept.1945

INSIGNIA — White clenched mail fist on black background

THEATRES — UK(10.40-11.42), N.Africa (11.42-3.44), Italy (3.44-5.45), Austria (5.45-8.45)

DIVISIONAL TROOPS — 1st Derbyshire Yeomanry (11.40-8.45), 51LAA.RA (6.42-11.44), 12RHA, 152Fld.RA, 72Atk.RA (6.42-8.45)

BRIGADES

20th ARMOURED BRIGADE (10.40-4.42)
1st Royal Glos.Hussars, 1st & 2nd Northamptonshire Yeomanries (10.40-4.42), 2Bn Rangers (11.40-3.41), 10Bn Kings Royal Rifle Corps (3.41-4.42)

26th ARMOURED BRIGADE (10.40-8.45)
16/5th Lancers, 17/21st Lancers, 2nd Lothian and Border Horse (10.40-8.45), 2Bn Tower Hamlet Rifles (10.40-1.41), 10Bn Rifle Brigade (2.41-4.44), 1Bn Kings Royal Rifle Corps (7.45-8.45)

38th INFANTRY BRIGADE (7.42-2.43)
2Bn London Irish Rifles, 1Bn Royal Irish Fusiliers, 6Bn Royal Inniskilling Fusiliers (7.42-2.43)

1st INFANTRY BRIGADE (GUARDS) (3.43-5.44)
3Bn Grenadier Guards, 2Bn Coldstream Guards, 3Bn Welsh Guards (3.43-5.44)

61st INFANTRY BRIGADE (6.44-8.45)
2Bn Rifle Brigade (6.44-8.45), 7Bn Rifle Brigade (6.44-5.45), 10Bn Rifle Brigade (6.44-2.45), 1Bn Kings Royal Rifle Corps (3.45-5.45), 2Bn Cameron Highlanders , 1Bn Welch Regiment, 1Bn Royal Sussex Regiment (6.45-8.45)

SUPPORT

6th SUPPORT GROUP (12.40-6.42)
9Bn Royal West Kent Regt (3.41-6.42), 51LAA.RA (1.41-5.42), 12RHA, 72Atk.RA (12.40-5.42)

7th ARMOURED DIVISION - Regular Division in Egypt, designated the Mobile Division to Sept.1939, then Armoured Division - Egypt, and from Feb.1940 the 7th Armoured Division. Still operational Sept.1945

INSIGNIA — White disc on red background, later red jerboa superimposed on disc. From 1944, brown jerboa, outlined white, on black.

THEATRES — N.Africa (9.39-9.43), Italy (9.43-12.43), UK (1.44-6.44), N.W.Europe (7.44-8.45)

DIVISIONAL TROOPS — 11th Hussars (4.40-11.43), 8th Hussars (12.43-8.45), 1RHA, 4RHA (8.42-9.42), 5RHA (12.42-8.45), 3RHA, 15LAA.RA (8.42-8.45), 74Fld.RA (9.42), 146Fld.RA (12.42-11.43), 4Fld.RA, 97Fld.RA (9.42-12.42), 65Atk.RA (9.42-5.45), 21Atk.RA (6.45-8.45)

BRIGADES

HEAVY ARMOURED BRIGADE - EGYPT (9.39-2.40) redesignated as :-
4th HEAVY ARMOURED BRIGADE (2.40-4.40) redesignated as :-
4th ARMOURED BRIGADE (4.40-7.41)(11.41-6.42)

1RTR (9.39-4.40), 6RTR (9.39-1.41), 2RTR (10.40-4.41) 3RTR, 8th Hussars (7.41)(11.41-6.42), 4RTR, 5RTR, 7RTR (5.41-7.41), 7th Hussars (5.40-4.41), 1Bn Kings Royal Rifle Corps (3.42-6.42)

LIGHT ARMOURED BRIGADE - EGYPT (9.39-2.40) redesignated as :-
7th LIGHT ARMOURED BRIGADE (2.40-4.40) redesignated as :-
7th ARMOURED BRIGADE (4.40-11.41)

7th Hussars (9.39-11.41), 8th Hussars (9.39-4.41), 11th Hussars (9.39-4.40), 3rd Hussars (9.40-1.41), 1RTR (5.40-4.41), 2RTR, 6RTR (5.41-11.41)

1st ARMY TANK BRIGADE (7.41-11.41)

8RTR, 42RTR, 44RTR (7.41-11.41)

22nd ARMOURED BRIGADE (7.42-8.45)

2nd Royal Glos. Hussars, 3rd County of London Yeomanry (7.42-9.42), 4th County of London Yeomanry (7.42-7.44), 1RTR, 5RTR (10.42-8.45), 5th Dragoon Guards (7.44-8.45), 1Bn Rifle Brigade (7.42-8.45)

contd....

7th ARMOURED DIVISION (contd)

BRIGADES (contd)

7th MOTOR BRIGADE (3.42-9.42)

1Bn & 9Bn Kings Royal Rifle Corps, 2Bn Rifle Brigade (3.42-6.42), 9Bn Kings Own Yorkshire Light Infantry, 1Bn Buffs, 14Bn Foresters (6.42-9.42)

131st INFANTRY BRIGADE (10.42-8.45)

1/5Bn Queens Royal Regiment(10.42-8.45), 1/6Bn & 1/7Bn Queens Royal Regiment (10.42-11.44), 9Bn Durham Light Infantry, 2Bn Devonshire Regiment (12.44-8.45)

SUPPORT

7th SUPPORT GROUP (2.40-3.42)

1Bn Kings Royal Rifle Corps (2.40-11.41), 2Bn Rifle Brigade (2.40-3.42), 3RHA/'M' Battery (2.40-2.41), 4RHA/'C' Battery (2.40-4.41), 1LAA(Bty).RA (2.41), 38LAA(Bty).RA (4.41-5.41), 1LAA.RA (5.41-1.42), 3RHA (3.41-2.42), 4RHA (4.41-2.42)

MACHINE GUN UNIT (2.44-8.45)

3rd Independent Machine Gun Company (2.44-8.45)

8th ARMOURED DIVISION - Formed in UK Dec.1940, disbanded in Egypt Jan.1943. Units of the Division were deployed on landing in N.Africa, usually serving with other Divisions. The 8th Armoured Division did not deploy as a complete unit whilst in N.Africa.

INSIGNIA The word 'GO' in black on a green disc within a black square.

THEATRES UK (12.40-5.42), N.Africa (7.42-12.42)

DIVISIONAL TROOPS 2nd Derbyshire Yeomanry (12.40-8.42), 104RHA (9.42), 5RHA (9.42-11.42), 11RHA (11.42), 146Fld.RA (9.42-11.42), 73Atk.RA (9.42-10.42), 56LAA.RA (7.42-11.42)

BRIGADES 23rd ARMOURED BRIGADE (12.40-6.42)

40RTR, 46RTR, 50RTR (12.40-6.42), 1Bn London Rifle Brigade (12.40-1.41), 7Bn Rifle Brigade (2.41-6.42)

24th ARMOURED BRIGADE (12.40-9.42)

41RTR, 45RTR, 47RTR (12.40-9.42), 1Bn Queens Westminsters (12.40-1.41), 11Bn Kings Royal Rifle Corps (2.41-9.42)

SUPPORT 8th SUPPORT GROUP (12.40- 7.42)

14Bn Foresters (12.40- 6.42), 73Atk.RA (12.40-7.42), 5RHA (12.40-7.42), 56LAA.RA (1.41-7.42)

9th ARMOURED DIVISION - Formed in UK Dec.1940, disbanded in UK August 1944

INSIGNIA	Panda's mask, no background specified
THEATRES	UK (12.40-8.44)
DIVISIONAL TROOPS	Inns of Court Regiment (1.41-1.43), 1st Royal Glos.Hussars (1.43-7.44), 6RHA (6.42-7.44), 141Fld.RA (6.42-7.44), 74Atk.RA (6.42-11.43), 92Atk.RA (11.43-7.44), 54LAA.RA (6.42-3.44), 150LAA.RA (3.44-7.44)
BRIGADES	27th ARMOURED BRIGADE (12.40-6.42) 4/7th Dragoon Guards, 13/18th Hussars, 1st East Riding Yeomanry (12.40-6.42), 1Bn Queen Victoria's Rifles (12.40-3.41), 7Bn Kings Royal Rifle Corps (4.41-6.42)
	28th ARMOURED BRIGADE (12.40-7.44) 5th Dragoon Guards, 15/19th Hussars, 1st Fife & Forfar Yeomanry (12.40-7.44), 2Bn Queen Victoria's Rifles (12.40-1.41), 8Bn Kings Royal Rifle Corps (2.41-7.44)
	7th INFANTRY BRIGADE (7.42-7.44) 2Bn South Wales Borderers (7.42-3.44), 6Bn Royal Sussex Regiment, 2/6Bn East Surrey Regiment (7.42-7.44)
SUPPORT	9th SUPPORT GROUP (12.40-6.42) 11Bn Worcestershire Regiment (12.40-6.42), 54LAA.RA (3.41-6.42), 74Atk.RA (12.40-6.42), 6RHA (12.40-6.42)

10th ARMOURED DIVISION   Formed by conversion and redesignation of the 1st Cavalry Division, Palestine July 1941. Disbanded in Egypt June 1944

INSIGNIA   Red foxes mask on yellow disc

THEATRES   Palestine (7.41-4.42), N.Africa (4.42-12.42), Palestine (1.43), Syria (1.43-9.43), N.Africa (9.43-6.44)

DIVISIONAL TROOPS   1st Household Cavalry Regiment (10.41-3.42), 2nd Derbyshire Yeomanry (8.42-9.42), Royals (9.42-10.42), 7th Hussars (11.43-4.44), 1RHA (9.42-5.43)(11.43-4.44), 14RHA (6.43-11.43), 104RHA (9.42-5.44), 98Fld.RA (9.42-12.42), 84Atk.RA (9.42-5.44), 53LAA.RA (9.42-11.42), 101LAA.RA (10.43-4.44)

BRIGADES   8th ARMOURED BRIGADE (7.41-11.42)

Greys (7.41-6.42), Staffordshire Yeomanry (7.41-11.42), Notts.(Sherwood Rangers) Yeomanry (7.41-11.42), 3RTR (6.42-11.42), 1Bn Buffs (3.42-11.42)

9th ARMOURED BRIGADE (10.41-3.42)(11.42-5.43)

Household Cavalry Regiment (10.41), Yorkshire Hussars (11.41-3.42), Royal Wiltshire Yeomanry, Warwickshire Yeomanry (10.41-3.42)(11.42.5.43), 3rd Hussars (11.42-5.43), 14Bn Foresters (11.42.12.42), 11Bn Kings Royal Rifle Corps (1.43-5.43)

7th ARMOURED BRIGADE (6.43-5.44)

7th Hussars (6.43-12.43), 2RTR, 6RTR (6.43-5.44), 8RTR (1.44-5.44), 2Bn Rifle Brigade (10.43-5.44)

23rd ARMOURED BRIGADE (6.44)

40RTR, 46RTR, 50RTR, 11Bn Kings Royal Rifle Corps (6.44)

133rd INFANTRY BRIGADE (10.42-11.42)

4Bn Buffs, 2Bn,4Bn & 5Bn Royal Sussex Regiment (10.42-11.42)

201st GUARDS MOTOR BRIGADE (11.42-2.43)

3Bn Coldstream Guards, 6Bn Grenadier Guards, 2Bn Scots Guards (11.42-2.43)

SUPPORT   No Support Group formed for this Division

<u>11th ARMOURED DIVISION</u> - Formed in UK Feb.1941, still operating Sept.1945

INSIGNIA — Black bull charging, red hooves, horns & eyes, rectangular yellow background

THEATRES — UK (3.41-6.44), N.W.Europe (6.44-8.45)

DIVISIONAL TROOPS — 27th Lancers (3.41-3.43), 2nd Northamptonshire Yeomanry (3.43-8.44), 15/19th Hussars (8.44-8.45), 13RHA (6.42-8.45), 151Fld.RA (6.42.8.45), 75Atk.RA (6.42-6.45), 65Atk.RA (6.45-8.45), 58LAA.RA (6.42-8.45)

BRIGADES — 29th ARMOURED BRIGADE (3.41-8.45)

23rd Hussars (3.41-8.45), 24th Lancers (3.41-2.44), 2nd Fife & Forfar Yeomanry (6.41-8.45), 3RTR (3.44-8.45), 8Bn Rifle Brigade (3.41-8.45)

30th ARMOURED BRIGADE (3.41-5.42)

22nd Dragoons, 1st Lothian & Border Horse, Westminster Dragoons, (3.41-5.42), 2Bn Queens Westminsters (3.41), 12Bn Kings Royal Rifle Corps (4.41-5.42)

159th INFANTRY BRIGADE (6.42-8.45)

4Bn Kings Shropshire Light Infantry, 3Bn Monmouthshire Regiment (6.42-8.45), 1Bn Herefordshire Regiment (6.42-3.45) 1Bn Cheshire Regiment (4.45-8.45)

SUPPORT — 11th SUPPORT GROUP (3.41-5.42)

8Bn Royal Ulster Rifles (3.41-5.41), 12Bn Green Howards (5.41-5.42), 13RHA, 75Atk.RA (3.41-5.42), 58LAA.RA (5.41-5.42)

MACHINE GUN UNIT (3.44-8.45)

2nd Independent Machine Gun Company (3.44-8.45)

<u>79th ARMOURED DIVISION</u> - Formed in UK, Oct.1942 as standard Armoured Division. Converted in June 1943 to Assault and Specialist Armour Division. Still operational Sept. 1945

INSIGNIA — Bulls head marked black and white, red and brown nostrils, red-tipped horns. Inverted yellow triangle background, head framed by inverted black triangle.

THEATRES — UK (10.42-6.44), N.W.Europe (6.44-8.45)

DIVISIONAL TROOPS — 162RAC (10.42-1.43), 2nd Northamptonshire Yeomanry (1.43-3.43), 142Fld.RA, 150Fld.RA, 55Atk.RA, 119LAA.RA (10.42-4.43)

43RTR (6.43-4.44), Staffordshire Yeomanry (9.44-4.45), 11RTR, 1st Fife & Forfar Yeomanry (9.44-12.44)

BRIGADES — 27th ARMOURED BRIGADE (10.42-5.43)
4/7th Dragoon Guards, 13/18th Hussars, 1st East Riding Yeomanry, 7Bn Kings Royal Rifle Corps (10.42-5.43)

185th INFANTRY BRIGADE (10.42-5.43)
2Bn Royal Warwickshire Regiment, 2Bn Royal Norfolk Regiment, 2Bn Kings Shropshire Light Infantry (10.42-5.43)

27th ARMOURED BRIGADE (6.43-12.43)
4/7th Dragoon Guards, 13/18th Hussars, 148RAC (6.43-12.43)

35th TANK BRIGADE (6.43-4.44)
49RTR,152RAC, 155RAC (6.43-4.44)

1st ASSAULT BRIGADE - ROYAL ENGINEERS (11.43-3.45) redesignated :-
1st ARMOURED ENGINEER BRIGADE (4.45-8.45)
5,6 & 42 Assault Regiments R.E. (11.43-3.45) redesignated :-
5,6 & 42 Armoured Engineer Regiments (4.45-8.45)

30th ARMOURED BRIGADE (1.44-6.45)
22nd Dragoons, 1st Lothian & Border Horse, Westminster Dragoons (1.44-6.45), 141RAC (6.44-9.44), 11RTR(12.44-4.45)

Contd...

79th ARMOURED DIVISION - Contd...

BRIGADES Contd..

1st TANK BRIGADE (5.44-8.44)

11RTR, 42RTR, 49RTR (5.44-8.44)

31st TANK BRIGADE (9.44-1.45) redesignated :-
31st ARMOURED BRIGADE (2.45-8.45)

141RAC (9.44-8.45), 11RTR(10.44-12.44), 7RTR (1.45-8.45), 1st Fife & Forfar Yeomanry (10.44-5.45), Greys (6.45-8.45), 49th Armoured Personnel Carrier Regiment (12.44-8.45)

33rd ARMOURED BRIGADE (1.45-8.45)

1st Northamptonshire Yeomanry (1.45-8.45), 1st East Riding Yeomanry (1.45-4.45), 144RAC (1.45-2.45), Staffordshire Yeomanry (5/45)

SUPPORT

No Support Group raised for this Division

42nd ARMOURED DIVISION - Formed by conversion and redesignation of 42nd Infantry Division, UK Nov.1941. Disbanded in UK October 1943

INSIGNIA Equi-sided white diamond within a red diamond

THEATRES UK (11.41-10.43)

DIVISIONAL TROOPS 112RAC (11.41-2.43), 1st Northamptonshire Yeomanry (1.43-3.43), 86Fld.RA (6.42-5.43), 147Fld.RA, 93LAA.RA (6.42-9.43), 191Fld.RA (5.43-10.43), 53Atk.RA (6.42-10.43),

BRIGADES 10th ARMOURED BRIGADE (11.41-5.42)
108RAC, 109RAC, 145RAC (11.41-5.42), 13Bn Highland Light Infantry (12.41-5.42)

11th ARMOURED BRIGADE (11.41-5.42)
107RAC, 110RAC, 111RAC (11.41-5.42), 1Bn Highland Light Infantry (12.41-5.42)

30th ARMOURED BRIGADE (6.42-10.43)
22nd Dragoons, 1st Lothian & Border Horse, Westminster Dragoons (6.42-10.43), 12Bn Kings Royal Rifle Corps (6.42-10.43)

71st INFANTRY BRIGADE (6.42-10.43)
1Bn East Lancashire Regiment, 1Bn Highland Light Infantry, 1Bn Oxfordshire & Buckinghamshire Light Infantry (6.42-10.43).

SUPPORT 42nd SUPPORT GROUP (12.41-5.42)
1Bn East Lancashire Regiment (12.41-5.42), 93LAA.RA (1.42-5.42), 147Fld.RA, 53Atk.RA (12.41-5.42)

GUARDS ARMOURED DIVISION - Formed in 1941 in UK by conversion and redesignation of 20th & 30th Infantry Brigades (Guards). Reverted to Guards Division (Infantry) June 1945.

INSIGNIA — Blue shield edged red with stylised white eye in centre

THEATRES — UK (9.41-6.44), N.W.Europe (8.44-6.45)

DIVISIONAL TROOPS — 2nd Household Cavalry Regiment (9.41-2.43), 2Bn Welsh Guards (4.43-6.45), 55Fld, 153Fld.RA (6.42-6.45), 21Atk.RA (6.42-5.45), 75Atk.RA (6.45), 94LAA.RA (6.42-6.45)

BRIGADES — 5th GUARDS ARMOURED BRIGADE (9.41-6.45)

2Bn Grenadier Guards, 1Bn Coldstream Guards, 1Bn Irish Guards (9.41-6.45), 1st Motorised Grenadier Guards (9.41-6.45)

6th GUARDS ARMOURED BRIGADE (9.41-2.43)

4Bn Grenadier Guards, 3Bn Scots Guards, 2Bn Welsh Guards (9.41-2.43), 4th Motorised Coldstream Guards (9.41-2.43)

32nd INFANTRY BRIGADE (GUARDS) (7.42-6.45)

5Bn Coldstream Guards (7.42-6.45), 4Bn Scots Guards (7.42-9.43), 1Bn Welsh Guards (7.42-2.45), 3Bn Irish Guards (9.43-2.45), 2Bn Scots Guards (2.45-6.45)

SUPPORT — GUARDS SUPPORT GROUP (9.41-6.42)

1Bn Welsh Guards (9.41-6.42), 153Fld.RA, 21Atk.RA (10.41-5.42), 94LAA.RA (1.42-5.42)

MACHINE GUN UNIT (3.44-6.45)

1st Independent Machine Gun Company (3.44-6.45)

1st INFANTRY DIVISION - Regular Division in the UK Sept.1939, still operating Sept.1945

INSIGNIA White equilateral triangle

THEATRES UK (9.39), France/Belgium (9.39-6.40), UK (6.40-2.43), N.Africa (3.43-12.43), Italy (12.43-1.45), Palestine (2.45-8.45)

DIVISIONAL TROOPS 13/18th Hussars (9.39-3.40), 1Bn-1Regt. Recce.Corps (1.41-12.43), 1st Recce.Regt.RAC (1.44-8.45), 2Fld.RA, 19Fld.RA (9.39-8.45), 24Fld.RA (9.39-1.40), 67Fld.RA (1.40-8.45), 21Atk.RA (9.39-10.41), 81Atk.RA (11.41-4.45), 93Atk.RA (4.45-8.45), 90LAA.RA (1.42-11.44), 11LAA.RA (11.44-1.45), 1LAA.RA (3.45-8.45)

BRIGADES 1st INFANTRY BRIGADE (GUARDS) (9.39-6.41)

3Bn Grenadier Guards, 2Bn Coldstream Guards (9.39-6.41), 2Bn Hampshire Regiment (9.39-5.41)

2nd INFANTRY BRIGADE (9.39-8.45)

1Bn Loyal Regiment, 2Bn North Staffordshire Regiment, (9.39-8.45), 1Bn Gordon Highlanders (9.39-2.40), 6Bn Gordon Highlanders (3.40-8.45)

3rd INFANTRY BRIGADE (9.39-8.45)

1Bn Duke of Wellingtons Regiment, 1Bn Kings Shropshire Light Infantry, 2Bn Foresters (9.39-8.45)

38th (IRISH) INFANTRY BRIGADE (2.42-6.42) redesignation of :-
210th INDEPENDENT INFANTRY BRIGADE (12.41-1.42)

2Bn London Irish Rifles, 1Bn Royal Irish Fusiliers, 6Bn Royal Inniskilling Fusiliers (12.41-6.42)

24th INFANTRY BRIGADE (GUARDS) (12.42-3.44)

1Bn Scots Guards, 1Bn Irish Guards, 5Bn Grenadier Guards (12.42-3.44)

18th INFANTRY BRIGADE (4.44-8.44)

1Bn Buffs, 9Bn Kings Own Yorkshire Light Infantry, 14Bn Foresters (4.44-8.44)

(contd...)

1st INFANTRY DIVISION (contd...)

BRIGADES (contd..)	66th INFANTRY BRIGADE (9.44-8.45) 2Bn Royal Scots, 1Bn Hertfordshire Regiment, 11Bn Lancashire Fusiliers (9.44-8.45) 34th TANK BRIGADE (7.42-9.42) Norther Irish Horse, 147RAC, 153RAC (7.42-9.42) 25th TANK BRIGADE (9.42-11.42) Northern Irish Horse, 51RTR, 142RAC (9.42-11.42)
SUPPORT	SUPPORT BATTALION (8.43-5.44) 2/7Bn Middlesex Regiment (8.43-5.44) MACHINE GUN UNITS (11.41-10.42)(5.44-8.45) 4Bn Cheshire Regiment (11.41-10.42), 2/7Bn Middlesex Regiment (5.44-8.45)

1st LONDON INFANTRY DIVISION - First line Territorial Division in UK, Sept.1939
Redesignated 56th London Infantry Division in Dec. 1940

INSIGNIA Black cat on red rectangular background

THEATRES UK (9.39-11.40)

DIVISIONAL TROOPS 1Bn Queen Victorias Rifles (M/Cycle) (9.39-5.40), 64Fld.RA (9.39-11.40), 90Fld.RA (9.39-11.40), 113Fld.RA (7.40-11.40), 67Atk.RA (7.40-11.40)

BRIGADES 1st LONDON INFANTRY BRIGADE (9.39-11.40)
8Bn Royal Fusiliers, 9Bn Royal Fusiliers, 7Bn Oxfordshire & Buckinghamshire Light Infantry (9.39-11.40)

2nd LONDON INFANTRY BRIGADE (9.39-11.40)
1Bn Queens Westminsters, 1Bn London Scottish, 10Bn Royal Berkshire Regiment (9.39-11.40)

35th INFANTRY BRIGADE (8.40-11.40)
2/5Bn, 2/6Bn & 2/7Bn Queens Royal Regiment (8.40-11.40)

SUPPORT No units

2nd INFANTRY DIVISION - Regular Infantry Division Sept.1939, still operating Sept.1945

INSIGNIA — Two white keys crossed on a black square

THEATRES — UK (9.39), France/Belgium (9.39-5.40), UK (6.40-4.42), India (6.42-4.44), Burma (4.44-4.45), India (4.45-8.45)

DIVISIONAL TROOPS — 4/7th Dragoon Guards (9.39-5.40), 2Bn-2Regt.Recce.Corps (4.41-12.43), 2nd Recce.Regt.RAC (1.44-8.45), 6Bn Loyal Regiment (M/Cycle) (11.40-4.41), 10Fld.RA (9.39-4.43) (7.43-8.45), 16Fld.RA (9.39-8.45), 18Fld.RA (9.39-2.40) 99Fld.RA (2.40-8.45), 13Atk.RA (9.39-7.42), 100Atk.RA (3.43-11.43)(10.44-8.45), 89LAA.RA (1.42-3.42), 60LAA.RA (4.43-7.43), 100LAA/Atk.RA (12.43-9.44)

BRIGADES

4th INFANTRY BRIGADE (9.39-8.45)

1Bn Royal Scots, 2Bn Royal Norfolk Regiment (9.39-8.45), 1Bn Border Regiment (9.39-4.40), 1/8Bn Lancashire Fusiliers (5.40-8.45)

5th INFANTRY BRIGADE (9.39-8.45)

2Bn Dorsetshire Regiment (9.39-8.45), 2Bn Royal Warwickshire Regiment, 1Bn Cameronians (9.39-1.40), 1Bn Cameron Highlanders, 7Bn Worcestershire Regiment (2.40-8.45)

6th INFANTRY BRIGADE (9.39-10.42)(7.43-8.45)

1Bn Royal Welsh Fusiliers, 1Bn Royal Berkshire Regiment (9.39-10.42)(7.43-8.45), 2Bn Durham Light Infantry (9.39-9.42) (7.43-8.45), 1Bn East Lancashire Regiment (9.41-10.41)

SUPPORT

MACHINE GUN UNIT (11.41-8.45)

2Bn Manchester Regiment (11.41-8.45)

2nd LONDON INFANTRY DIVISION - Second line Territorial Division, Sept.1939, redesignated 47th London Infantry Division in Dec.1940

INSIGNIA Two red bells with red bow on black background

THEATRES UK (9.39-11.40)

DIVISIONAL TROOPS 2Bn Queen Victorias Rifles (M/Cycle) (9.39-6.40), 117Fld.RA, 138Fld.RA (9.39-11.40), 114Fld.RA (7.40-11.40), 62Atk.RA (10.39-11.40)

BRIGADES 4th LONDON INFANTRY BRIGADE (9.39-11.40)

11Bn Royal Fusiliers, 12Bn Royal Fusiliers, 2Bn London Irish Rifles (9.39-11.40)

5th LONDON INFANTRY BRIGADE (9.39-11.40)

2Bn Queens Westminsters, 2Bn London Scottish, 2Bn London Rifle Brigade (9.39-11.40)

25th INFANTRY BRIGADE (6.40-11.40)

2Bn Essex Regiment, 4Bn Border Regiment (6.40-11.40), 4Bn Buffs (6.40-10.40)

SUPPORT No units

3rd INFANTRY DIVISION - Regular Division in UK Sept.1939, still operating Sept.1945

INSIGNIA — Red inverted triangle surrounded by three black triangles, the whole forming one equilateral triangle

THEATRES — UK (9.39), France/Belgium (9.39-5.40), UK (6.40-6.44), N.W.Europe (6.44-8.45)

DIVISIONAL TROOPS — 8Bn Royal Northumberland Fusiliers (M/Cycle) (11.40-4.41), 15/19th Hussars (9.39-3.40), 3Bn-3Regt. Recce.Corps (4.41-12.43), 3rd Recce.Regt. RAC (1.44-8.45), 7Fld.RA, 33Fld.RA (9.39-8.45), 23Fld.RA (9.39-3.40), 76Fld.RA (3.40-8.45), 6Fld.RA (8.45), 20Atk.RA (9.39-8.45), 92LAA.RA (3.42-6.45), 112LAA.RA (6.45-8.45)

BRIGADES

7th INFANTRY BRIGADE (GUARDS) (9.39-9.41)

1Bn & 2Bn Grenadier Guards, 1Bn Coldstream Guards (9.39-9.41)

7th INFANTRY BRIGADE (12.41-5.42)

2Bn South Wales Borderers, 6Bn Royal Sussex Regiment, 2/6Bn East Surrey Regiment (12.41-5.42)

8th INFANTRY BRIGADE (9.39-8.45)

1Bn Suffolk Regiment, 2Bn East Yorkshire Regiment (9.39-8.45), 2Bn Gloucestershire Regiment (9.39-1.40), 4Bn Royal Berkshire Regiment (2.40-6.40), 1Bn South Lancashire Regiment (7.40-8.45)

9th INFANTRY BRIGADE (9.39-8.45)

2Bn Lincolnshire Regiment, 1Bn Kings Own Scottish Borderers, 2Bn Royal Ulster Rifles (9.39-8.45)

185th INFANTRY BRIGADE (5.43-8.45)

2Bn Royal Warwickshire Regiment, ~~2Bn~~ 1BN Royal Norfolk Regiment, 2Bn Kings Shropshire Light Infantry (5.43-8.45)

33rd TANK BRIGADE (6.42-4.43)

43RTR, 144RAC, 148RAC (6.42-4.43)

SUPPORT

MACHINE GUN BATTALION (11.41-5.42)(3.44-8.45)
SUPPORT BATTALION (10.43-3.44)

2Bn Middlesex Regiment

4th INFANTRY DIVISION - Regular Division in UK, Sept.1939. Still operating Sept.1945

INSIGNIA Fourth quadrant of circle in red, later red disc with fourth quadrant protruding on white background

THEATRES UK (9.39), France/Belgium (10.39-6.40), UK (6.40-3.43), N.Africa (3.43-2.44), Italy (2.44-12.44), Greece (12.44-8.45)

DIVISIONAL TROOPS 5th Dragoon Guards (9.39-3.40), 4Bn-4Regt. Recce. Corps (1.41-12.43), 4th Recce.Regt. RAC (1.44-8.45), 22Fld.RA, 30Fld.RA (9.39-8.45), 17Fld.RA (9.39.2.40), 77Fld.RA (2.40-8.45), 14Atk.RA (9.39-8.45), 91LAA.RA (1.42-11.44)

BRIGADES 10th INFANTRY BRIGADE (9.39-8.45)

2Bn Bedfordshire & Hertfordshire Regiment, 2Bn Duke of Cornwalls Light Infantry (9.39-8.45), 1Bn Royal West Kent Regiment (9.39-4.40), 1/6Bn East Surrey Regiment (5.40-8.45)

11th INFANTRY BRIGADE (9.39-6.42)

2Bn Lancashire Fusiliers, 1Bn East Surrey Regiment (9.39-6.42), 1Bn Oxfordshire & Buckinghamshire Light Infantry (9.39-1.40), 5Bn Northamptonshire Regiment (2.40-6.42)

12th INFANTRY BRIGADE (9.39-8.45)

2Bn Royal Fusiliers (9.39-8.45), 1Bn South Lancashire Regiment (9.39-6.40), 1Bn Black Watch (9.39-2.40), 6Bn Black Watch (3.40-8.45), 1Bn Royal West Kent Regiment (9.40-8.45)

28th INFANTRY BRIGADE (1.44-8.45)

2Bn Kings Regiment, 2Bn Somerset Light Infantry (1.44-8.45), 2/4Bn Hampshire Regiment (4.44-8.45)

21st TANK BRIGADE (6.42-12.43)

12RTR, 48RTR, 145RAC (6.42-12.43)

SUPPORT MACHINE GUN UNIT (11.41-5.42)(6.44-8.45)
SUPPORT BATTALION (3.44-6.44)

2Bn Royal Northumberland Fusiliers

5th INFANTRY DIVISION - Regular Division in UK, Sept.1939, still operating Sept.1945

INSIGNIA White 'Y' on Khaki background

THEATRES UK (9.39-12.39), France/Belgium (12.39-5.40), UK (6.40-3.42), India (5.42-8.42), Iraq (8.42-9.42), Persia (9.42-1.43), Syria (2.42-6.43), N.Africa (6.43), Sicily (7.43-9.43), Italy (9.43-7.44), Palestine (7.44-2.45), Italy (2.45), N.W.Europe (3.45-8.45)

* 13th & 17th Brigades detatched 5.42-7.42 to serve with Force 121 in Madagascar

DIVISIONAL TROOPS 3Bn Tower Hamlets Rifles (1.41), 5Bn-5Regt. Recce.Corps (1.41-12.43), 5th Recce.Regt.RAC (1.44-8.45), 9Fld.RA (12.39-4.42), 91Fld.RA (12.39-4.42)(7.42-8.45), 92Fld.RA (12.39-4.40)(6.40-6.45), 97 Army Fld.RA (5.40), 156Fld.RA (8.42-8.45), 52Atk.RA (12.39-8.45), 18LAA.RA (2.43-8.45)

BRIGADES 13th INFANTRY BRIGADE (9.39-8.45)*

2Bn Cameronians, 2Bn Royal Inniskilling Fusiliers (9.39-8.45), 2Bn Wiltshire Regiment (9.39-8.44), 5Bn Essex Regiment (9.44-8.45)

15th INFANTRY BRIGADE (9.39-8.45)

1Bn Green Howards, 1Bn Kings Own Yorkshire Light Infantry, 1Bn York & Lancaster Regiment (9.39-8.45)

17th INFANTRY BRIGADE (9.39-8.45)*

2Bn Royal Scots Fusiliers, 2Bn Northamptonshire Regiment (9.39-8.45), 2Bn Seaforth Highlanders (9.39-3.40), 6Bn Seaforth Highlanders (4.40-8.45)

SUPPORT MACHINE GUN UNIT (11.41-8.45)

7Bn Cheshire Regiment (11.41-8.45)

6th INFANTRY DIVISION - Formed Sept.1939 by redesignation of 7th Infantry Division. Redesignated HQ Western Desert from July 1940 to Feb. 1941 when the Division re-formed. Redesignated 70th Infantry Division, Sept.1941.

INSIGNIA Red four-pointed star on white background

THEATRES N.Africa (9.39-3.40), Palestine (3.40-6.40), N.Africa (6.40-6.41), Syria (6.41-9.41)

DIVISIONAL TROOPS Greys (3.40-5.40), 60Fld.RA (7.41-9.41)

BRIGADES

22nd INFANTRY BRIGADE (9.39-3.40)
2Bn Scots Guards (9.39-3.40), 1Bn Buffs (9.39-11.39), 1Bn Welch Regiment (12.39-3.40)

14th INFANTRY BRIGADE (4.40-6.40)(5.41-9.41)
2Bn York & Lancaster Regiment, 2Bn Kings Own Royal Regiment, 1Bn South Staffordshire Regiment (4.40-6.40), 2Bn Black Watch, 2Bn York & Lancaster Regiment (5.41-9.41), 1Bn Welch Regiment (5.41), 1Bn Bedfordshire & Hertfordshire Regiment (6.41-9.41)

16th INFANTRY BRIGADE (4.40-6.40)(3.41-9.41)
2Bn Leicestershire Regiment, 2Bn Queens Royal Regiment (4.40-6.40)(3.41-9.41), 1Bn Argyll & Sutherland Highlanders (4.40-6.40)(3.41-5.41), 2Bn Kings Own Royal Regiment (6.41-9.41)

22nd GUARDS BRIGADE (3.41-4.41)
3Bn Coldstream Guards, 2Bn Scots Guards, 1Bn Durham Light Infantry (3.41-4.41)

23rd INFANTRY BRIGADE (5.41-9.41)
4Bn Border Regiment, 1Bn Buffs, 1Bn Durham Light Infantry (5.41-9.41)

SUPPORT No units

8th INFANTRY DIVISION - Pre War Division in Palestine, disbanded Feb.1940

INSIGNIA	Red cross of St. George on blue shield
THEATRES	Palestine (9.39-2.40)
DIVISIONAL TROOPS	No units
BRIGADES	14th INFANTRY BRIGADE (9.39-2.40) 2Bn Queens Royal Regiment, 1Bn Argyll & Sutherland Highlanders (9.39-1.40), 2Bn Rifle Brigade (9.39-10.39), 1Bn Welch Regiment (11.39), 1Bn Bedfordshire & Hertfordshire Regiment (11.39-1.40), 2Bn York & Lancaster Regiment (1.40-2.40) 16th INFANTRY BRIGADE (9.39-2.40) 2Bn Leicestershire Regiment (9.39-2.40), 1Bn South Staffordshire Regiment (9.39-10.39), 1Bn Welch Regiment (9.39-11.39), 1Bn Foresters (9.39-1.40), 2Bn Rifle Brigade (10.39-1.40), 2Bn West Yorkshire Regiment (10.39-12.39), 1Bn Buffs (12.39-1.40), 1Bn Argyll & Sutherland Highlanders (1.40-2.40), 2Bn Queens Royal Regiment (1.40-2.40)
SUPPORT	No units

9th (HIGHLAND) INFANTRY DIVISION - Second line Territorial Division Sept.1939, redesignated 51st Highland Division Aug.1940

INSIGNIA — Silver thistle on dark blue background within silver circular border.

THEATRES — UK (9.39-8.40)

DIVISIONAL TROOPS — 126Fld.RA, 127Fld.RA, 128Fld.RA, 61Atk.RA (9.39-8.40)

BRIGADES — 26th INFANTRY BRIGADE (9.39-8.40)

5Bn & 7Bn Seaforth Highlanders, 5Bn Cameron Highlanders (9.39-8.40)

27th INFANTRY BRIGADE (9.39-8.40)

5Bn Black Watch (9.39-8.40), 7Bn Gordon Highlanders (9.39-11.39), 9Bn Gordon Highlanders (12.39-8.40)

28th INFANTRY BRIGADE (9.39-8.40)

10Bn & 11Bn Argyll & Sutherland Highlanders (9.39-8.40), 7Bn Black Watch (9.39-6.40)

SUPPORT — No units

12th (EASTERN), INFANTRY DIVISION - Second line Territorial Division Sept.1939
Disbanded July 1940 in UK

INSIGNIA White diamond

THEATRES UK (10.39-4.40), France (4.40-6.40), UK (6.40-7.40)

DIVISIONAL TROOPS 113Fld.RA, 114Fld.RA, 118Fld.RA, 67Atk.RA (10.39-7.40)
* above units did not accompany the Division to France

BRIGADES 35th INFANTRY BRIGADE (9.39-7.40)
2/5Bn, 2/6Bn & 2/7Bn Queens Royal Regiment (9.39-7.40)

36th INFANTRY BRIGADE (9.39-7.40)
6Bn & 7Bn Royal West Kent Regiment, 5Bn Buffs (9.39-7.40)

37th INFANTRY BRIGADE (9.39-7.40)
6Bn & 7Bn Royal Sussex Regiment, 2/6Bn East Surrey Regiment (9.39-7.40)

SUPPORT No units

15th (SCOTTISH) INFANTRY DIVISION - Second line Territorial Division Sept.1939, still Operating Sept.1945

INSIGNIA: Scottish Lion rampant on yellow disc, bordered white, on a black square

THEATRES: UK (9.39-6.44), N.W.Europe (6.44-8.45)

DIVISIONAL TROOPS: 15Bn. Recce.Corps (1.41-12.41), 15Indpdt.Coy-Sqdn. Recce.Corps (12.41-2.43), 15Regt. Recce.Corps (2.43-12.43), 15 Recce. Regt. RAC (1.44-8.45), 129Fld.RA (9.39-5.42), 130Fld.RA (9.39-1.42), 131Fld.RA (9.39-8.45), 181Fld.RA (11.42-8.45), 190Fld.RA (3.43-8.45), 64Atk.RA (9.39-3.42), 97Atk.RA (8.42-12.44), 119LAA.RA (5.43-8.45)

BRIGADES:

44th INFANTRY BRIGADE (9.39-8.45)
8Bn Royal Scots, 6Bn Kings Own Scottish Borderers (9.39-8.45), 7Bn Kings Own Scottish Borderers (9.39-10.42), 11Bn Argyll & Sutherland Highlanders (10.42-12.42), 6Bn Royal Scots Fusiliers (1.43-8.45)

45th INFANTRY BRIGADE (9.39-1.43)
9Bn & 10Bn Cameronians (9.39-12.42), 6Bn Royal Scots Fusiliers (9.39-12.39)(7.40-12.42), 10Bn Cameronians, 10Bn Black Watch, 11Bn Argyll & Sutherland Highlanders (1.43)

46th INFANTRY BRIGADE (9.39-8.45)
2Bn Glasgow Highlanders (9.39-8.45), 10Bn & 11Bn Highland Light Infantry (9.39-11.41), 7Bn Cameronians (12.41-3.42), 7Bn Seaforth Highlanders (12.41-8.45), 4Bn Cameron Highlanders (4.42-11.42), 9Bn Cameronians (12.42-8.45)

227th INFANTRY BRIGADE (10.43-8.45)
10Bn Highland Light Infantry, 2Bn Argyll & Sutherland Highlanders, 2Bn Gordon Highlanders (10.43-8.45)

6th GUARDS TANK BRIGADE (2.43-9.43)
4Bn Grenadier Guards, 4Bn Coldstream Guards, 3Bn Scots Guards (2.43-9.43)

SUPPORT: MACHINE GUN UNIT (3.44-8.45), SUPPORT BATTALION (10.43-3.44)
1Bn Middlesex Regiment

18th INFANTRY DIVISION - Second line Territorial Division in UK Sept.1939. Captured by Japanese in Malaya Feb.1942, not re-formed

INSIGNIA — Black diamond overset by black diagonal cross, orange background

THEATRES — UK (9.39-10.41), India (12.41-1.42), Malaya(Singapore) (1.42-2.42)

DIVISIONAL TROOPS — 5Bn Loyal Regiment (M/Cycle) (1.41-4.41), 18Bn Recce.Corps (4.41-2.42), 105Fld.RA (9.39-11.39), 135Fld.RA (9.39-2.42), 148Fld.RA (9.39-2.42), 118Fld.RA (6.40-2.42), 65Atk.RA (9.39-11.39), 125Atk.RA (7.40-2.42)

BRIGADES — 53rd INFANTRY BRIGADE (9.39-2.42)

5Bn & 6Bn Royal Norfolk Regiment, 2Bn Cambridgeshire Regiment (9.39-2.42)

54th INFANTRY BRIGADE (9.39-2.42)

4Bn Royal Norfolk Regiment, 4Bn & 5Bn Suffolk Regiment (9.39.2.42)

55th INFANTRY BRIGADE (9.39-2.42)

5Bn Bedfordshire & Hertfordshire Regiment, 1/5Bn Foresters, 1Bn Cambridgeshire Regiment (9.39-2.42)

SUPPORT — No units

23rd (NORTHUMBRIAN) INFANTRY DIVISION - Second line Territorial Division, Sept.1939. Disbanded in the UK, June 1940

INSIGNIA	Tudor rose in white set on blue (sometimes green) background
THEATRES	UK (9.39-4.40), France (4.40-6.40), UK (6.40)
DIVISIONAL TROOPS	124Fld.RA, 125Fld.RA (10.39-6.40), 8Bn Northumberland Fusiliers (M/Cycle) (10.39-6.40)
BRIGADES	69th INFANTRY BRIGADE (9.39-6.40) 6Bn & 7Bn Green Howards, 5Bn East Yorkshire Regiment (9.39-6.40) 70th INFANTRY BRIGADE (9.39-6.40) 10Bn & 11Bn Durham Light Infantry (9.39-6.40), 12Bn Durham Light Infantry (9.39-1.40), 1Bn Tyneside Scottish (2.40-6.40)
SUPPORT	MACHINE GUN UNIT (10.39-6.40) 9Bn Northumberland Fusiliers (10.39-6.40)

<u>36th INFANTRY DIVISION</u> - Formed by redesignation of 36th Indian Infantry Division in Burma, Sept.1944. Still operating Sept.1945

INSIGNIA	White and red circles interlinked on black background
THEATRES	Burma (9.44-5.45), India (5.45-8.45)
DIVISIONAL TROOPS	2Bn Border Regiment (4.45-8.45), 130 Aslt.Fld.RA (9.44-6.45), 178 Aslt.Fld.RA (9.44-7.45), 145Fld.RA (6.45-8.45), 8Fld.RA, 158Fld.RA (7.45-8.45), 455Lt.Bty.RA (9.44-11.44), 122Atk.RA (9.44-8.45), 122LAA/Atk.RA (9.44)
BRIGADES	26th INFANTRY BRIGADE (4.45-8.45) 2Bn Buffs, 1Bn Devonshire Regiment, 1Bn Northamptonshire Regiment (4.45-8.45) 29th INFANTRY BRIGADE (9.44-8.45) 1Bn Royal Scots Fusiliers, 2Bn Royal Welsh Fusiliers, 2Bn East Lancashire Regiment (9.44-5.45), 2Bn Queens Royal Regiment, 1Bn Essex Regiment, 1Bn Cambridgeshire Regiment (6.45-8.45) 72nd INFANTRY BRIGADE (9.44-8.45) 6Bn South Wales Borderers, 10Bn Gloucestershire Regiment, 9Bn East Sussex Regiment (9.44-8.45)
SUPPORT	MACHINE GUN UNIT (9.44-4.45) 'D' Coy. 2Bn Manchester Regiment DEFENCE BATTALION (5.45-8.45) 2Bn Leicestershire Regiment (5.45-8.45)
INDIAN ARMY	1Bn 7th Rajput Regiment (MG) (6.45-8.45), 12Mtn.Bty.IA, 28Mtn.Bty.IA (12.44-2.45), 32Mtn.IA (2.45-8.45) 26th Indian Infantry Brigade 12.44-4.45 when it was redesignated 26th Infantry Brigade

38th (WELSH) INFANTRY DIVISION - Second line Territorial Division Sept.1939. Ceased to command units Aug.1944 until its redesignation as reserve Division Sept.1944. Still operating Sept.1945

INSIGNIA Yellow cross of St.David on black shield, khaki background

THEATRES UK (9.39-8.45)

DIVISIONAL TROOPS 38Indpt.Coy.-Sqdn. Recce.Corps (1.42-10.43), 161Regt.Recce.Corps (10.43-11.43), 38Regt.Recce.Corps (11.43-12.43), 38th Recce.Regt. RAC (1.44-8.44)(10.44-11.44), 80th Recce.Regt. RAC (9.44-10.44), 102Fld.RA (9.39-11.41),132Fld.RA (9.39-7.42), 146Fld.RA (9.39-5.42), 183Fld.RA (7.42-4.43), 182Fld.RA (1.43-8.44), 193Fld.RA (5.43-10.43), 195Fld.RA (9.44-8.45), 70Atk.RA (10.39-7.44)(9.44-8.45)

BRIGADES 113th INFANTRY BRIGADE (9.39-1.44)(10.44-8.45)

15Bn & 2/5Bn Welch Regiment (9.39-1.44), 1Bn South Wales Borderers (12.42-1.44), 4Bn Monmouthshire Regiment (9.39-11.42), 4Bn East Lancashire Regiment, 5Bn Border Regiment, 5Bn Royal Inniskilling Fusiliers (10.44-8.45)

114th INFANTRY BRIGADE (9.39-8.45) - less Aug.-Sept.1944

5Bn Kings Shropshire Light Infantry (9.39-8.45), Brecknock Bn. - South Wales Borderers (9.39-8.44), 2Bn Herefordshire Regiment (9.39-8.44), 9Bn Royal Welsh Fusiliers, 8Bn Royal Warwickshire Regiment, 8Bn Worcestershire Regiment (10.44-8.45)

115th INFANTRY BRIGADE (9.39-7.44)

8Bn Royal Welsh Fusiliers (9.39-7.44), 9Bn Royal Welsh Fusiliers (9.39-10.43), 10Bn Royal Welsh Fusiliers (9.39-8.42), 13Bn Royal Welsh Fusiliers (8.42-7.44), 9Bn Somerset Light Infantry (11.43-7.44), 30Bn Royal Berkshire Regiment (8.44-3.45)

SUPPORT MACHINE GUN UNIT (9.44-8.45)

5Bn Cheshire Regiment (9.44-8.45)

42nd (EAST LANCASHIRE) INFANTRY DIVISION - First line Territorial Division Sept.1939. Converted & redesignated 42nd Armoured Division Nov.1941

INSIGNIA: Equi-sided white diamond within a red diamond

THEATRES: UK (9.39-4.40), France/Belgium (4.40-5.40), UK (6.40-10.41)

DIVISIONAL TROOPS: 51Fld.RA (9.39-4.40), 52Fld.RA (9.39-10.41), 53Fld.RA (9.39-10.41), 111Fld.RA (7.40-10.41), 147Fld.RA (10.41), 56Atk.RA (9.39-10.41), 53Atk.RA (10.41)

BRIGADES:

125th INFANTRY BRIGADE (9.39-11.41)
1/5Bn & 1/6Bn Lancashire Fusiliers (9.39-11.41), 1/8Bn Lancashire Fusiliers (9.39-4.40), 1Bn Border Regiment (5.40-11.40), 9Bn Lancashire Fusiliers (12.40-11.41)

126th INFANTRY BRIGADE (9.39-11.41)
5Bn Kings Own Royal Regiment, 5Bn Border Regiment (9.39-11.41), 1Bn East Lancashire Regiment (9.39-7.41), 5Bn Manchester Regiment (8.41-11.41)

127th INFANTRY BRIGADE (9.39-7.41)
5Bn Manchester Regiment (9.39-7.41), 5Bn East Lancashire Regiment, 8Bn Manchester Regiment (9.39-5.40), 4Bn East Lancashire Regiment, 1Bn Highland Light Infantry (6.40-7.41)

SUPPORT: No units

43rd (WESSEX) INFANTRY DIVISION - First line Territorial Division Sept.1939. Still operating Sept.1945

INSIGNIA Heraldic Wyvern in gold on dark blue square

THEATRES UK (9.39-8.45)

DIVISIONAL TROOPS 48Bn.Recce.Corps (11.41-1.42), 43Bn-43Regt.Recce.Corps (1.42-12.43), 43rd Recce.Regt. RAC (1.44-8.45), 94Fld.RA, 112Fld.RA (9.39-6.45), 141Fld.RA (9.39-6.42), 179Fld.RA (6.42-8.45), 59Atk.RA (9.39-8.45), 110LAA.RA (3.42-8.45)

BRIGADES

128th INFANTRY BRIGADE (9.39-6.42)
1/4Bn, 2/4Bn & 5Bn Hampshire Regiment (9.39-6.42)

129th INFANTRY BRIGADE (9.39-8.45)
4Bn & 5Bn Wiltshire Regiment, 4Bn Somerset Light Infantry (9.39-8.45)

130th INFANTRY BRIGADE (9.39-8.45)
4Bn & 5Bn Dorsetshire Regiment, 7Bn Hampshire Regiment (9.39-8.45)

214th INFANTRY BRIGADE (10.43-8.45)
9Bn Somerset Light Infantry (10.43), 7Bn Somerset Light Infantry, 5Bn Duke of Cornwalls Light Infantry (10.43-8.45), 1Bn Worcestershire Regiment (11.43-8.45)

25th TANK BRIGADE (7.42-9.42)
11RTR (7.42), 51RTR, 142RAC (7.42-9.42), Northern Irish Horse (9.42)

34th TANK BRIGADE (10.42-9.43)
147RAC, 151RAC, 153RAC (10.42-9.43)

SUPPORT

MACHINE GUN UNIT (11.41-10.42)(2.44-8.45)
1/8Bn Middlesex Regiment (11.40-10.42), 8Bn Middlesex Regiment (2.44-8.45)

SUPPORT BATTALION (7.43-3.44)
8Bn Middlesex Regiment (10.43-2.44)

44th (HOME COUNTIES) INFANTRY DIVISION - First line Territorial Division Sept.1939. Disbanded in N.Africa Jan. 1943

INSIGNIA Red oval, sometimes outlined in white

THEATRES UK (9.39-3.40), France/Belgium (4.40-5.40), UK (6.40-5.42), N.Africa (7.42-1.43)

DIVISIONAL TROOPS 44Bn-44Regt. Recce.Corps (1.41-11.42), 57Fld.RA, 58Fld.RA, 65Fld.RA (9.39-11.42), 57Atk.RA (9.39-11.42), 99LAA.RA (2.42-3.42), 30LAA.RA (4.42-11.42)

BRIGADES 131st INFANTRY BRIGADE (9.39-9.42)

1/7Bn Queens Royal Regiment (9.39-5.40)(7.41-9.42), 1/5Bn & 1/6Bn Queens Royal Regiment (9.39-9.42), 2Bn Buffs (6.40-7.41)

132nd INFANTRY BRIGADE (9.39-9.42)

4Bn & 5Bn Royal West Kent Regiment (9.39-9.42), 2Bn Buffs (7.41-9.42), 1/6Bn East Surrey Regiment (9.39-5.40), 1/7Bn Queens Royal Regiment (6.40-6.41)

133rd INFANTRY BRIGADE (9.39-6.42)

2Bn, 5Bn & 4Bn Royal Sussex Regiment, 2Bn Buffs (9.39-6.42)

SUPPORT MACHINE GUN UNIT (11.41-11.42)

6Bn Cheshire Regiment (11.41-11.42)

45th INFANTRY DIVISION - Second line Territorial Division Sept.1939. Redesignated 45th (Holding) Division from Oct.1944 for reception & training of personnel returning from overseas. Finally designated 45th Division Dec.1944-Aug.1945

INSIGNIA Yellow drum, red bands top and bottom, white cords and small red diamond and dark blue square in centre. Khaki background

THEATRES UK (9.39-8.45)

DIVISIONAL TROOPS 45Bn.Recce.Regt. (1.41-12.41), 45Indpdt.Coy-Sqdn. Recce.Corps (1.42-12.43), 161st Recce.Regt. RAC (9.44-8.45), 55Fld.RA (9.39-6.42), 96Fld.RA (9.39-6.44), 142Fld.RA (9.39-11.41), 171Fld.RA (2.42-8.44), 175Fld.RA, 176Fld.RA (9.44-8.45), 69Atk.RA (9.39-6.42), 90Atk.RA (2.42-11.43), 88Atk.RA (1.44-8.44), 8AA.RA-8.RA (Reserve) (9.44-8.45)

BRIGADES 134th INFANTRY BRIGADE (9.39-8.44)(9.44-8.45)

6Bn Devonshire Regiment (9.39-8.44), 8Bn Devonshire Regiment (9.39-1.43), 4Bn Devonshire Regiment (9.39-5.40), 1Bn Royal Irish Fusiliers (6.40-10.40), 9Bn Devonshire Regiment (11.40-9.42), 9Bn Dorsetshire Regiment (10.42-11.43), 2Bn East Surrey Regiment (2.43-8.44), 4Bn Devonshire Regiment (11.43-3.44)

* Brigade disbanded 8.44 & reformed from 203 Inf. Bde. 9.44

7Bn Royal Ulster Rifles, 11Bn Argyll & Sutherland Highlanders, 2/6Bn Lancashire Fusiliers, (9.44-8.45), 14Bn Durham Light Infantry (11.44-8.45)

135th INFANTRY BRIGADE (9.39-7.44)(9.44-8.45)

5Bn & 6Bn Somerset Light Infantry (9.39-7.44), 7Bn Somerset Light Infantry (9.39-9.42), 7Bn Wiltshire Regiment (10.42-3.44), 4Bn Devonshire Regiment (4.44-7.44)

** Brigade disbanded 7.44 & reformed from 209 Inf. Bde. 9.44

14Bn Durham Light Infantry (9.44-11.44), 11Bn South Staffordshire Regiment (11.44-8.45), 11BnHampshire Regiment, 2/5Bn Welch Regiment, 6Bn Northamptonshire Regiment (9.44-8.45)

(Contd...)

45th INFANTRY DIVISION (Contd..)

BRIGADES (Contd..)

136th INFANTRY BRIGADE (9.39-7.44)(9.44-8.45)

4Bn Duke of Cornwalls Light Infantry (9.39-7.44), 5Bn Duke of Cornwalls Light Infantry (9.39-8.42), 9Bn Devonshire Regiment (9.39-11.40), 1Bn Royal Irish Fusiliers (12.40-11.41), 14Bn Foresters (12.41-4.42), 2Bn Kings Shropshire Light Infantry (4.42-8.42), 7Bn Royal West Kent Regiment (8.42-1.43), 12Bn Hampshire Regiment (8.42-7.44), 10Bn East Surrey Regiment (2.43-7.44)

** Brigade disbanded 7.44, new HQ formed 9.44 for reception, selection & training duties.

137th INFANTRY BRIGADE (3.45-8.45)

Resurected Brigade, original having been converted to the 137th Armoured Brigade 7.42. New Brigade HQ formed for reception, selection & training duties

178th INFANTRY BRIGADE (5.45-8.45)

HQ formed for reception, selection & training duties

SUPPORT

No units

46th INFANTRY DIVISION - Second line Territorial Division Sept.1939, still operating Sept.1945

INSIGNIA — Oak tree, brown trunk & green foliage, picked out by narrow white border, black square background

THEATRES — UK (10.39-4.40), France/Belgium (4.40-6.40), UK (6.40-1.43), N.Africa (1.43-9.43), Italy (9.43-3.44), Palestine (4.44-6.44), Italy (7.44-1.45), Greece (1.45-4.45), Italy (4.45-5.45), Austria (5.45-8.45)

DIVISIONAL TROOPS — 46Bn-46Regt. Recce.Corps (7.41-12.43), 46th Recce.Regt. RAC (1.44-8.45), 121Fld.RA, 122Fld.RA, 123Fld.RA (10.39-7.40), 70Fld.RA, 71Fld.RA (7.40-8.45), 51Fld.RA (7.40-9.40), 151Fld.RA (9.40-5.42), 172Fld.RA (5.42-8.45), 68Atk.RA (10.39-6.40), 58Atk.RA (7.40-8.45), 115LAA.RA (2.42-11.44)

BRIGADES

137th INFANTRY BRIGADE (9.39-7.42)
2/6Bn & 2/7Bn Duke of Wellingtons Regiment, 2/5Bn West Yorkshire Regiment (9.39-7.42)

138th INFANTRY BRIGADE (9.39-8.45)
6Bn Lincolnshire Regiment, 6Bn York & Lancaster Regiment, 2/4Bn Kings Own Yorkshire Light Infantry (9.39-8.45)

139th INFANTRY BRIGADE (9.39-8.45)
2/5Bn Leicestershire Regiment (9.39-8.45), 2/5Bn Foresters (9.39-2.43), 9Bn Foresters (9.39-12.40), 5Bn Foresters (3.43-11.44)(3.45-8.45), 16Bn Durham Light Infantry (1.41-8.45)

128th INFANTRY BRIGADE (8.42-8.45)
1/4Bn & 2/4Bn Hampshire Regiment (8.42-8.45), 5Bn Hampshire Regiment (8.42-4.43), 2Bn Hampshire Regiment (5.43-8.45)

SUPPORT

MACHINE GUN UNITS (11.41-10.42)(7.44-8.45)
2/7Bn Middlesex Regiment (11.41-10.42), 9Bn Manchester Regiment (7.44-8.45)

SUPPORT BATTALION (7.43-3.44)
2Bn Northumberland Fusiliers (7.43-3.44)

47th (LONDON) INFANTRY DIVISION - Formed by redesignation of 2nd London Infantry Division, Dec.1940, still operating Sept.1945

INSIGNIA Two red bells with red bow on black background

THEATRES UK (12.40-8.45)

DIVISIONAL TROOPS 47th Indpdt.Coy-Sqdn. Recce.Corps (12.41-6.43), 117Fld.RA (12.40-8.44)(11.44-8.45), 138Fld.RA (12.40-7.42), 114Fld.RA (12.40-11.41), 75Fld.RA (2.43-4.43), 187Fld.RA (4.43-6.43), 5Fld.RA (6.43-7.44), 168Fld.RA (6.43-11.43), 173Fld.RA (9.44-12.44), 171Fld.RA (5.45-8.45), 62Atk.RA (12.40-2.43), 78Atk.RA (2.43-11.43), 89Atk.RA (11.43-8.44), 56Med.RA (9.44-8.45), 171Hvy.RA (2.45-5.45)

BRIGADES 25th INFANTRY BRIGADE (12.40-7.44)

2Bn Essex Regiment (12.40-12.41)(10.42-2.44), 4Bn Border Regiment (12.40-1.41), 9Bn Kings Own Royal Regiment (12.40-9.42), 12Bn South Staffordshire Regiment (2.41-11.41), 19Bn & 20Bn Royal Fusiliers (12.41-9.42), 11Bn York & Lancaster Regiment, 6Bn Kings Own Yorkshire Light Infantry (10.42-12.42), 1Bn Duke of Cornwalls Light Infantry, 9Bn Royal Norfolk Regiment (1.43-7.44), 7Bn North Staffordshire Regiment (3.44-7.44)

140th LONDON INFANTRY BRIGADE (12.40-7.44)

11Bn Royal Fusiliers (12.40-7.44), 12Bn Royal Fusiliers (12.40-8.43), 2Bn London Irish Rifles (12.40-11.41), 6Bn Oxfordshire & Buckinghamshire Light Infantry (12.41-3.42), 6Bn Royal Inniskilling Fusiliers (4.42-9.42), 17Bn Royal Fusiliers (9.42-12.42), 2Bn Gloucestershire Regiment (1.43-2.44), 2/4Bn Essex Regiment (9.43-7.44), 11Bn West Yorkshire Regiment (3.44-7.44)

141st LONDON INFANTRY BRIGADE (12.40-7.44)

16Bn & 17Bn Royal Fusiliers (12.40-10.42), 2Bn London Scottish (12.40-7.44), 6Bn Royal Inniskilling Fusiliers (11.42-4.44), 8Bn Black Watch (11.42-10.43), 4Bn Black Watch (11.43-7.44), 2Bn Royal Inniskilling Fusiliers (4.44-7.44)

Contd...

47th (LONDON) INFANTRY DIVISION (Contd..)

BRIGADES (Contd..)

7th INFANTRY BRIGADE (8.44-8.45)
13Bn Queens Royal Regiment, 12Bn Royal Fusiliers, 2/6Bn East Surrey Regiment (10.44-8.45).
(No Battalions under command 8.44-10.44)

140th INFANTRY BRIGADE (8.44-8.45)
4Bn Oxfordshire & Buckinghamshire Light Infantry, 6Bn Royal Sussex Regiment, 7Bn Gloucestershire Regiment (8.44-8.45)

141st INFANTRY BRIGADE (12.44-8.45)
8Bn Suffolk Regiment, 9Bn Dorsetshire Regiment, 6Bn Devonshire Regiment (12.44-8.45)

220th INFANTRY BRIGADE (8.44-11.44)
5Bn West Yorkshire Regiment, 9Bn Royal Welsh Fusiliers, 2/6Bn Lancashire Fusiliers, 6Bn Kings Own Royal Regiment (8.44-11.44)

SUPPORT No support units allocated

48th (SOUTH MIDLAND) INFANTRY DIVISION - First line Territorial Division Sept.1939, still operating Sept.1945

INSIGNIA — Blue macaw set on red diamond within a dark blue oval

THEATRES — UK (9.39-1.40), France/Belgium (1.40-5.40), UK (6.40-8.45)

DIVISIONAL TROOPS — 1st Lothian & Border Horse (9.39-4.40), 5Bn Gloucestershire Regiment (7.41-10.41), 48Bn. Recce.Corps (10.41-11.41), 48Indpdt.Coy-Sqdn. Recce.Corps (1.42-1.43), 67Fld.RA (9.39-1.40), 68Fld.RA (9.39-8.42), 99Fld.RA (9.39-2.40), 18Fld.RA (2.40-2.41), 24Fld.RA (1.40-8.42), 23Fld.RA (2.41-11.41), 173Fld.RA (1.42-3.42), 180Fld.RA (12.42-8.44), 120Fld.RA (9.44-8.45), 53Atk.RA (9.39-10.41), 93Atk.RA (11.41-8.42), 96Atk.RA (9.42-8.45), 168Med.RA (4.45-6.45), 12Med.RA (7.45-8.45)

BRIGADES

143rd INFANTRY BRIGADE (9.39-8.45)

1/7Bn Royal Warwickshire Regiment (9.39-9.42), 8Bn Royal Warwickshire Regiment (9.39-7.44), 5Bn Northamptonshire Regiment (9.39-1.40), 1Bn Oxfordshire & Buckinghamshire Light Infantry (2.40-10.41), 10Bn Somerset Light Infantry (11.41-9.42), 8Bn Suffolk Regiment (10.42-7.44), 11Bn Green Howards (10.42-7.44), 10Bn Duke of Wellingtons Regiment, 11Bn York & Lancaster Regiment, 11Bn West Yorkshire Regiment (8.44-8.45)

144th INFANTRY BRIGADE (9.39-8.45)

5Bn Gloucestershire Regiment (9.39-4.41), 7Bn Worcestershire Regiment (9.39-1.41), 8Bn Worcestershire Regiment (9.39-7.44), 2Bn Royal Warwickshire Regiment (3.40-11.40), 11Bn South Staffordshire Regiment (11.40-1.43), 9Bn Somerset Light Infantry (5.41-9.42), 7Bn Royal Ulster Rifles (9.42-10.43), 17Bn Royal Fusiliers (2.43-10.43), 4Bn Oxfordshire & Buckinghamshire Light Infantry,(5Bn Royal Inniskilling Fusiliers (11.43-7.44), 10Bn Cameronians, 9Bn Seaforth Highlanders, 10Bn Black Watch, 2Bn London Scottish (8.44-8.45)

145th INFANTRY BRIGADE (9.39-11.43)

4Bn Oxfordshire & Buckinghamshire Light Infantry (9.39-11.43) and Buckinghamshire Bn. (9.39-12.42), 4Bn Royal Berkshire Regiment (9.39-2.40), 2Bn Gloucestershire Regiment (3.40-12.42), 6Bn Kings Own Royal Regiment, 5Bn Royal Inniskilling Fusiliers (1.43-11.43)

(NO SUPPORT UNITS)

49th (WEST RIDING) INFANTRY DIVISION - First line Territorial Division Sept.1939 Division suspended operations 4.40 when HQ became HQ 'AVONFORCE' (Norway 5.40) & HQ 'ALABASTERFORCE' (Iceland 7.40). Division reconstituted 4.42, still operating Sept.1945

INSIGNIA · White polar bear, head down in earlier form, head up later form. Black background.

THEATRES UK (9.39-4.40)(8.42-6.44), N.W.Europe (6.44-8.45)

DIVISIONAL TROOPS 49Regt. Recce.Corps (9.42-12.43), 49th Recce.Regt. RAC (1.44-8.45), 69Fld.RA (9.39-6.40)(9.42-8.45), 70Fld.RA, 71Fld.RA (9.39-6.40), 79Fld.RA, 80Fld.RA (6.40),143Fld.RA (4.42-8.45), 178Fld.RA (5.42-12.42), 185Fld.RA (12.42-11.44), 74Fld.RA (11.44-8.45), 58Atk.RA (9.39-6.40), 88Atk.RA (6.42-11.43), 55Atk.RA (7.43-8.45), 118LAA.RA (5.42-8.42), 89LAA.RA (2.42-8.45)

BRIGADES 146th INFANTRY BRIGADE (9.39-4.40)(8.42-8.45)

4Bn Lincolnshire Regiment, 1/4Bn Kings Own Yorkshire Light Infantry, Hallamshire Bn. York & Lancaster Regiment (9.39-4.40)(8.42-8.45)

147th INFANTRY BRIGADE (9.39-4.40)(4.42-8.45)

1/5Bn West Yorkshire Regiment (9.39-4.40)(4.42-8.42), 1/6Bn & 1/7Bn Duke of Wellingtons Regiment (9.39-4.40)(4.42-1.43), 11Bn Royal Scots Fusiliers (9.42-8.45), 6Bn Duke of Wellingtons Regiment (2.43-7.44), 7Bn Duke of Wellingtons Regiment (2.43-8.45), 1Bn Leicestershire Regiment (8.44-8.45)

148th INFANTRY BRIGADE (9.39-4.40)

1/5Bn Leicestershire Regiment, 2Bn South Wales Borderers, 8Bn Foresters (9.39-4.40)

70th INFANTRY BRIGADE (5.42-8.44)

10Bn & 11Bn Durham Light Infantry, 1Bn Tyneside Scottish (5.42-8.44)

56th INFANTRY BRIGADE (8.44-8.45)

2Bn South Wales Borderers, 2Bn Gloucestershire Regiment, 2Bn Essex Regiment (8.44-8.45)

SUPPORT MACHINE GUN UNIT (2.44-8.45) SUPPORT BATTALION (6.43-2.44)

2Bn Kensington Regiment (2.44-8.45) (6.43-2.44)

50th (NORTHUMBRIAN) INFANTRY DIVISION - First line Territorial Division Sept.1939, still operating Sept.1945

INSIGNIA — Two linked 'T's in red on a black square

THEATRES — UK (9.39-1.40), France/Belgium (1.40-6.40), UK (6.40-4.41), N.Africa (6.41-7.41), Cyprus (7.41-11.41), Iraq (11.41-1.42), Syria (1.42-2.42), N.Africa (2.42-6.43), Sicily (7.43-10.43), UK (11.43-6.44), N.W.Europe (6.44-12.44), UK (12.44-8.45), Norway (8.45)

DIVISIONAL TROOPS — 4Bn Royal Northumberland Fusiliers (M/Cycle) (9.39-6.40), 50Bn Recce.Corps (3.41-11.41), Yorkshire Hussars (12.43-1.44), 61st Recce.Regt. RAC (1.44-11.44), 72Fld.RA (9.39-11.41)(2.42-6.42), 74Fld.RA (9.39-11.44), 124Fld.RA (6.40-12.44), 111Fld.RA (10.42-11.42), 65Fld.RA (11.42-4.43), 90Fld.RA (4.43-11.44), 65Atk.RA (11.39-11.40), 102Atk.RA (10.42-11.44), 34LAA.RA 10.42-11.42), 25LAA.RA (12.42-11.44)

BRIGADES — 150th INFANTRY BRIGADE (9.39-6.41)(9.41-12.41)(4.42-6.42)

4Bn East Yorkshire Regiment, 4Bn & 5Bn Green Howards (9.39-6.41)(9.41-12.41)(4.42-6.42)

151st INFANTRY BRIGADE (9.39-8.45)

6Bn & 8Bn Durham Light Infantry (9.39-8.45), 9Bn Durham Light Infantry (9.39-11.44), 1/7Bn Queens Royal Regiment (12.44-8.45)

69th INFANTRY BRIGADE (7.40-8.45)

5Bn East Yorkshire Regiment, 6Bn & 7Bn Green Howards (7.40-8.45)

168th LONDON INFANTRY BRIGADE (5.43-10.43)

1Bn London Irish Rifles, 1Bn London Scottish, 10Bn Royal Berkshire Regiment (5.43-10.43)

231st INFANTRY BRIGADE (11.43-8.45)

1Bn Dorsetshire Regiment, 1Bn Hampshire Regiment (11.43-8.45), 2Bn Devonshire Regiment (11.43-11.44), 1/6Bn Queens Royal Regiment (12.44-8.45)

SUPPORT — MACHINE GUN UNIT (2.41-6.45)

2Bn Cheshire Regiment (2.41-6.45)

51st (HIGHLAND) INFANTRY DIVISION - First line Territorial Division, Sept.1939 Captured in France, June 1940. Reconstituted in UK Sept.1940 from 9th Highland Division, still operating Sept.1945

INSIGNIA Letters 'HD' in red linked together within a red circle on a blue background.

THEATRES UK (9.39-1.40), France (1.40-6.40), UK (8.40-6.42), N.Africa (8.42-7.43), Sicily (7.43-11.43), UK (11.43-6.44), N.W.Europe (6.43-8.45)

DIVISIONAL TROOPS 1st Fife & Forfar Yeomanry (9.39-3.40), 51Bn-51Regt. Recce.Corps (1.41-11.42), 2nd Derbyshire Yeomanry (1.44-8.45), 75Fld.RA (9.39-6.40), 76Fld.RA (9.39-3.40), 77Fld.RA (9.39-2.40), 17Fld.RA (2.40-6.40), 23Fld.RA (3.40-6.40), 126Fld.RA, 127Fld.RA, 128Fld.RA (8.40-8.45), 51Atk.RA (9.39-6.40), 61Fld.RA (8.40-8.45), 40LAA.RA (5.42-8.45)

BRIGADES 152nd INFANTRY BRIGADE (9.39-6.40)(9.40-8.45)

4Bn Seaforth Highlanders, 4Bn Cameron Highlanders (9.39-6.40), 6Bn Seaforth Higlanders (9.39-3.40), 2Bn Seaforth Highlanders (4.40-6.40) RECONSTITUTED... 2Bn Seaforth Highlanders, 5Bn Cameron Highlanders (9.40-8.45), 4/5Bn Seaforth Highlanders (9.40-4.41), 5Bn Seaforth Highlanders (5.41-8.45)

153rd INFANTRY BRIGADE (9.39-6.40)(9.40-8.45)

4Bn Black Watch, 5Bn Gordon Highlanders (9.39-6.40), 6Bn Gordon Highlanders (9.39-3.40), 1Bn Gordon Highlanders (4.40-6.40) RECONSTITUTED... 1Bn Gordon Highlanders, 5/7Bn Gordon Highlanders, 5Bn Black Watch (9.40-8.45)

154th INFANTRY BRIGADE (9.39-6.40)(9.40-8.45)

7Bn & 8Bn Argyll & Sutherland Highlanders (9.39-6.40), 6Bn Black Watch (9.39-3.40), 1Bn Black Watch (4.40-6.40) RECONSTITUTED... 1Bn Black Watch (9.40-8.45), 11Bn Argyll & Sutherland Highlanders (9.40-9.41), 7/10Bn Argyll & Sutherland Highlanders (9.40-10.42), 7Bn Black Watch (10.41-8.45), 7Bn Argyll & Sutherland Highlanders (11.42-8.45)

SUPPORT MACHINE GUN UNIT (11.41-12.43)(2.44-8.45)

1/7Bn Middlesex Regiment (11.41-12.43)(2.44-8.45)

SUPPORT BATTALION (12.43-2.44)

1/7Bn Middlesex Regiment (12.43-2.44)

52nd (LOWLAND) INFANTRY DIVISION - First line Territorial Division Sept.1939, Still operating Sept.1945

INSIGNIA: Cross of St.Andrew on blue shield, Khaki background. Scroll with 'MOUNTAIN' added circa 1942

THEATRES: UK (9.39-6.40), France (6.40), UK (6.40-10.44), N.W.Europe (10.44-8.45)

DIVISIONAL TROOPS: 52Bn-52Regt. Recce.Corps (1.41-12.43), 52nd Recce.Regt. RAC (1.44-8.44), 78Fld.RA (9.39-6.42), 79Fld.RA (9.39-8.45), 80Fld.RA (9.39-8.45), 186Fld.RA (12.42-8.45), 54Atk.RA (9.39-8.45), 108LAA.RA (3.42-8.45)

BRIGADES: 155th INFANTRY BRIGADE (9.39-8.45)
7/9Bn Royal Scots, 4Bn Kings Own Scottish Borderers (9.39-8.45), 5Bn Kings Own Scottish Borderers (9.39-2.45), 6Bn Highland Light Infantry (3.45-8.45)

156th INFANTRY BRIGADE (9.39-8.45)
4/5Bn Royal Scots Fusiliers, 6Bn Cameronians (9.39-8.45), 7Bn Cameronians (9.39-2.45), 1Bn Glasgow Highlanders (3.45-8.45)

157th INFANTRY BRIGADE (9.39-8.45) No battalions 9.44-11.44)
5Bn Highland Light Infantry (9.39-8.45), 6Bn Highland Light Infantry, 1Bn Glasgow Hihlanders (9.39-8.44), 5Bn Kings Own Scottish Borderers, 7Bn Cameronians (12.44-8.45)

SUPPORT: MACHINE GUN UNITS (11.41-5.42)(3.44-8.45)
1/9Bn Manchester Regiment (11.41-5.42), 7Bn Manchester Regiment (3.44-8.45)

SUPPORT BATTALION (6.43-3.44)
7Bn Manchester Regiment (6.43-3.44)

<u>53rd (WELSH) INFANTRY DIVISION</u> - First line Territorial Division Sept.1939, still operating Sept.1945

INSIGNIA Red 'W', centre vee infilled, standing on red bar. Black background

THEATRES UK (9.39-6.44), N.W.Europe (6.44-8.45)

DIVISIONAL TROOPS 53Bn-53Regt. Recce.Corps (1.41-12.43), 53rd Recce.Regt. RAC (1.44-8.45), 81Fld.RA, 83Fld.RA, 133Fld.RA (9.39-8.45), 71Atk.RA (10.40-4.41)(6.41-8.45), 63Atk.RA (4.41-6.41), 116LAA.RA (4.42-12.44), 25LAA.RA (12.44-8.45)

BRIGADES

158th INFANTRY BRIGADE (9.39-8.45)
4Bn & 6Bn Royal Welsh Fusiliers (9.39-8.44), 7Bn Royal Welsh Fusiliers (9.39-4.45), 1Bn East Lancashire Regiment, 1/5Bn Welch Regiment (9.44-8.45), 2Bn South Wales Borderers (5.45-8.45)

159th INFANTRY BRIGADE (9.39-5.42)
4Bn Kings Shropshire Light Infantry, 3Bn Monmouthshire Regiment, 1Bn Herefordshire Regiment (9.39-5.42)

160th INFANTRY BRIGADE (9.39-8.45)
4Bn Welch Regiment, 2Bn Monmouthshire Regiment (9.39-8.45), 1/5Bn Welch Regiment (9.39-8.44), 6Bn Royal Welsh Fusiliers (9.44-8.45)

71st INFANTRY BRIGADE (10.43-8.45)
1Bn Highland Light Infantry, 1Bn Oxfordshire & Buckinghamshire Light Infantry (10.43-8.45), 1Bn East Lancashire Regiment (10.43-8.44), 4Bn Royal Welsh Fusiliers (9.44-8.45)

31st TANK BRIGADE (6.42-9.43)
9RTR (6.42-9.43), 10RTR (6.42-1.43) 7RTR (2.43-9.43)

SUPPORT

MACHINE GUN UNIT (11.41-12.42)(2.44-8.45)
5Bn Cheshire Regiment (11.41-12.42), 1Bn Manchester Regiment (2.44-8.45)

SUPPORT BATTALION (10.43-2.44)
1Bn Manchester Regiment (10.43-2.44)

54th (EAST ANGLIAN) INFANTRY DIVISION - First line Territorial Division Sept.1939
Division ceased operating Dec.1943 with
HQ redesignated HQ(L.of C)-54 Division

INSIGNIA Letters 'JP' with common vertical in blue on a red disc

THEATRES UK (9.39-12.43)

DIVISIONAL TROOPS 54Bn. Recce.Corps (7.41-12.41), 54Indpdt.Coy.-Sqdn. Recce.Corps (1.42-2.43), 85Fld.RA (9.39-8.42), 86Fld.RA (9.39-6.42), 134Fld.RA (9.39-2.42), 168Fld.RA (6.42-6.43), 55Atk.RA (9.39-2.43), 199LAA.RA (4.42-9.42)

BRIGADES 161st INFANTRY BRIGADE (9.39-11.40)
1/5Bn & 2/5Bn Essex Regiment (9.39-11.40), 1/4Bn Essex Regiment (9.39-7.40)

162nd INFANTRY BRIGADE (9.39-11.42)(9.43-12.43)
1Bn & 2Bn Hertfordshire Regiment (9.39-9.42), 6Bn Bedfordshire & Hertfordshire Regiment (9.39-11.42)(9.43-12.43), 16Bn Foresters (10.42-11.42), 7Bn East Yorkshire Regiment (10.42-11.42)(9.43-12.43), 1Bn Leicestershire Regiment (9.43-12.43)

163rd INFANTRY BRIGADE (9.39-1 .43)
5Bn Royal Berkshire Regiment (9.39-11.43), 2/4Bn Essex Regiment (9.39-5.43), 7Bn Royal Berkshire Regiment (9.39-9.42), 6Bn Kings Own Royal Regiment (10.42-12.42), 1 Buckinghamshire Bn. Oxfordshire & Buckinghamshire Light Infantry (1.43-5.43), 5Bn Kings Regiment (6.43-11.43)

198th INFANTRY BRIGADE (12.40-8.43)
8Bn Kings Regiment, 6Bn Border Regiment (12.40-8.43), 7Bn Border Regiment (9.39-11.42), 2Bn Hertfordshire Regiment (12.42-8.43)

SUPPORT No units

55th (WEST LANCASHIRE) INFANTRY DIVISION - First line Territorial Division Sept.1939, still operating Sept.1945

INSIGNIA Red rose of Lancashire on khaki disc

THEATRES UK (9.39-8.45)

DIVISIONAL TROOPS 5Bn Loyal Regiment (M/Cycle) (9.39-4.40), 55Indpdt.Coy.-Sqdn. Recce.Corps (12.41-10.43), 161st Recce.Regt. RAC (1.44-7.44), 1st Royal Glos. Hussars (8.44-6.45), 161Regt.Recce.Corps. (12.43), 87Fld.RA (9.39-11.41), 136Fld.RA (9.39-8.42), 109Fld.RA (7.40-8.45), 174Fld.RA (7.42-3.43), 192Fld.RA (2.43-12.43), 170Fld.RA (5.43-3.44), 5Fld.RA (8.44-6.45), 5Lt.RA (6.45-8.45), 141Fld.RA (8.44-6.45), 124Fld.RA (6.45-8.45), 89Atk.RA (10.44-8.45), 66Atk.RA (9.39-11.39)(7.40-11.44), 103LAA.RA (2.42-11.42), 149LAA.RA (8.43-12.43), 150LAA.RA (8.44-2.45)

BRIGADES 164th INFANTRY BRIGADE (9.39-5.45)

9Bn Kings Regiment, 2/4Bn South Lancashire Regiment (9.39-942), 1/4Bn South Lancashire Regiment (9.39-8.44), 17Bn Durham Light Infantry (10.42-8.43), 10Bn Green Howards (10.42-4.43), 9Bn South Lancashire Regiment, 9Bn Buffs (9.43-8.44), 1Bn Duke of Cornwalls Light Infantry, 4Bn Devonshire Regiment, 5Bn Somerset Light Infantry (9.44-5.45)

165th INFANTRY BRIGADE (9.39-8.45)

2Bn Liverpool Scottish (9.39-8.42), 1Bn Liverpool Scottish (9.39-7.44), 5Bn Kings Regiment (9.39-4.43), 10Bn Duke of Wellingtons Regiment (9.42-8.44), 9Bn Kings Regiment (5.43-7.44), 4Bn Black Watch, 2Bn Royal Irish Fusiliers, 5Bn West Yorkshire Regiment (8.44-8.45)

199th INFANTRY BRIGADE (9.39-8.44) redesignated :-
166th INFANTRY BRIGADE (9.44-8.45)

2/8Bn Lancashire Fusiliers (9.39-8.44), 6Bn Manchester Regiment (9.39-4.42), 7Bn Manchester Regiment (9.39-8.42), 1Bn Manchester Regiment (5.42-8.42), 2Bn Loyal Regiment (9.42-11.44), 5Bn Royal Inniskilling Fusiliers (9.42-12.42), 11Bn South Staffordshire Regiment (1.43-10.43), 9Bn Bedfordshire & Hertfordshire Regiment (11.43-7.44), 1Bn Liverpool Scottish, 1/4Bn South Lancashire Regiment (8.44-8.45), 8Bn Manchester Regiment (12.44-8.45)

SUPPORT MACHINE GUN UNIT - 5Bn Manchester Regiment (10.44-8.45)

56th (LONDON) INFANTRY DIVISION - Formed by redesignation of 1st London Infantry Division, Dec.1940. Still operating Sept.1945

INSIGNIA	Black cat on red rectangular background
THEATRES	UK (12.40-8.42), Iraq (11.42-3.43), Palestine (3.43), N.Africa (3.43-8.43)(4.44-7.44), Italy (9.43-3.44)(7.44-8.45)
DIVISIONAL TROOPS	56Bn-56Regt. Recce.Corps (1.41-8.42), 44Regt. Recce.Corps (3.43-12.43), 44Recce.Regt RAC (1.44-2.44)(6.44-8.45), 64Fld.RA (12.40-8.45), 90Fld.RA (12.40-3.43), 113Fld.RA (12.40-8.45), 65Fld.RA (4.43-8.45), 67Atk.RA (12.40-8.45), 100LAA.RA (2.42-11.44)
BRIGADES	167th LONDON INFANTRY BRIGADE (12.40-8.45) 8Bn Royal Fusiliers (12.40-9.44), 9Bn Royal Fusiliers (12.40-8.45), 15Bn Royal Fusiliers (12.40-3.41), 7Bn Oxfordshire & Buckinghamshire Light Infantry (3.41-9.44), 1Bn London Irish Rifles (10.44-8.45), 1Bn London Scottish (10.44-8.45) 169th LONDON INFANTRY BRIGADE (12.40-8.45) 2/5Bn, 2/6Bn & 2/7Bn Queens Royal Regiment (12.40-8.45) 168th LONDON INFANTRY BRIGADE (12.40-4.43)(11.43-9.44) 18Bn Royal Fusiliers (12.40-3.41), 1Bn London Irish Rifles (12.40-4.43)(11.43-9.44), 1Bn London Scottish (12.40-4.43)(11.43-9.44), 10Bn Royal Berkshire Regiment (4.41-4.43)(11.43-5.44), 1Bn Welch Regiment (6.44-9.44) 201st GUARDS MOTOR BRIGADE (8.43-10.43) 3Bn Coldstream Guards, 6Bn Grenadier Guards (8.43-10.43) 24th INFANTRY BRIGADE (GUARDS) (4.45-8.45) 2Bn Coldstream Guards, 1Bn Scots Guards, 1Bn Buffs (4.45-8.45)
SUPPORT	MACHINE GUN UNITS (1.43-8.45)(11.41-5.42) 6Bn Cheshire Regiment (1.43-8.45), 1Bn Kensington Regiment (1.41-5.42)

59th (STAFFORDSHIRE) INFANTRY DIVISION - Second line Territorial Division Sept.1939, Operations suspended Aug.1944

INSIGNIA Pit-head gear in red against a black slag heap, blue background

THEATRES UK (9.39-8.44)

DIVISIONAL TROOPS 6Bn Loyal Regiment (M/Cycle) (9.39-6.40), 59Bn-59Regt.Recce.Corps (1.41-12.43), 59th Recce.Regt. RAC (1.44-8.44), 61Fld.RA, 116Fld.RA (9.39-8.44), 110Fld.RA (7.40-8.44), 66Atk.RA (11.39-6.40), 68Atk.RA (7.40-8.44), 68LAA.RA (4.43-8.44)

BRIGADES 176th INFANTRY BRIGADE (9.39-8.44)

7Bn South Staffordshire Regiment, 6Bn North Staffordshire Regiment (9.39-8.44), 7Bn North Staffordshire Regiment (9.39-10.42), 7Bn Royal Norfolk Regiment (11.42-8.44)

177th INFANTRY BRIGADE (9.39-8.44)

5Bn, 1/6Bn & 2/6Bn South Staffordshire Regiment (9.39-8.44)

197th INFANTRY BRIGADE (7.40-8.44)

2/5Bn Lancashire Fusiliers, 5Bn East Lancashire Regiment (7.40-8.44), 2/6Bn Lancashire Fusiliers (7.40-10.42), 1/7Bn Royal Warwickshire Regiment (11.42-8.44)

SUPPORT MACHINE GUN UNIT (11.41-10.42)(2.44-8.44)

7Bn Royal Northumberland Fusiliers (11.41-10.42)(2.44-8.44)

SUPPORT BATTALION (10.43-2.44)

7Bn Royal Northumberland Fusiliers (10.43-2.44)

61st INFANTRY DIVISION - Second line Territorial Division Sept.1939, still operating Sept.1945

INSIGNIA Red diamond on blue background

THEATRES UK (9.39-8.45)

DIVISIONAL TROOPS 61Bn-61Regt. Recce.Corps (9.41-12.43), 61st Recce.Regt. RAC (1.44), Yorkshire Hussars (1.44-6.45), 2Bn Kings Royal Rifle Corps (7.45-8.45), 199Fld.RA (9.39-8.45), 120Fld.RA (9.39-6.44), 145Fld.RA (9.39-1.44)(8.44-1.45), 96Fld.RA (6.44-1.45), 141Fld.RA (6.45-8.45), 168Lt.RA (6.45-8.45), 71Atk.RA (4.41-6.41), 63Atk.RA (9.39-7.40)(6.41-8.44), 92Atk.RA (10.44-6.45), 103LAA.RA (2.43-3.44), 148LAA.RA (3.44-8.45)

BRIGADES

182nd INFANTRY BRIGADE (9.39-8.45)

2/7Bn Royal Warwickshire Regiment, 9Bn Worcestershire Regiment (9.39-8.45), 9Bn Royal Warwickshire Regiment (9.39-9.44), 1Bn South Wales Borderers (10.44-8.45)

183rd INFANTRY BRIGADE (9.39-3.44)(7.44-8.45)

7Bn Gloucestershire Regiment, 10Bn Worcestershire Regiment, 4Bn Northamptonshire Regiment (9.39-3.44), 5Bn Somerset Light Infantry, 4Bn Devonshire Regiment, 1Bn Duke of Cornwalls Light Infantry (5.45-8.45), 7Bn North Staffordshire Regiment, 1Bn Foresters (7.44-4.45)

184th INFANTRY DIVISION (9.39-3.44)(7.44-8.45)

5Bn Oxfordshire & Buckinghamshire Light Infantry (9.39-3.44), 2nd Buckinghamshire Bn. Oxfordshire & Buckinghamshire Light Infantry (9.39-3.44), 6Bn Royal Berkshire Regiment (9.39-3.44) (7.44-5.45), 2Bn Royal West Kent Regiment, 2Bn East Surrey Regiment (7.44-8.45), 1Bn Foresters (6.45-8.45)

SUPPORT

MACHINE GUN UNITS (11.41-5.42)(7.44-9.44)(7.45-8.45)

2/8Bn Middlesex Regiment (11.41-5.42), 4Bn Cheshire Regiment (7.44-9.44), 2Bn Cheshire Regiment (7.45-8.45)

SUPPORT BATTALION (10.43-9.44)

4Bn Cheshire Regiment (10.43-9.44)

66th INFANTRY DIVISION - Second line Territorial Division Sept.1939. Disbanded June 1940 in the UK

INSIGNIA Light blue equilateral triangle divided horizontally by a yellow band, no background specified. (This design was adopted but never issued)

THEATRE UK (9.39-6.40)

DIVISIONAL TROOPS 109Fld.RA, 110Fld.RA, 111Fld.RA (9.39.6.40)

BRIGADES 197th INFANTRY BRIGADE (9.39-6.40)

2/5Bn & 2/6Bn Lancashire Fusiliers, 5Bn East Lancashire Regiment (9.39-6.40)

198th INFANTRY BRIGADE (9.39-6.40)

6Bn & 7Bn Border Regiment, 8Bn Kings Regiment (9.39-6.40)

199th INFANTRY BRIGADE (9.39-6.40)

6Bn & 7Bn Manchester Regiment, 2/8Bn Lancashire Fusiliers (9.39-6.40)

SUPPORT No units

70th INFANTRY DIVISION - Formed in Syria Sept.1941 by redesignation of 6th Infantry Division. Disbanded in India Nov.1943

INSIGNIA	Red four-pointed star on white background
THEATRES	Syria (9.41-10.41), N.Africa (10.41-2.42), Syria-Egypt (2.42), India (3.42-10.43)
DIVISIONAL TROOPS	45Bn Recce.Corps (10.42-9.43), 8Fld.RA (2.42-9.43), 51Fld.RA (3.42)(2.43-9.43), 60Fld.RA (12.41-10.43), 56Atk.RA (4.42-7.43), 69LAA/Atk.RA (8.43-9.43)
BRIGADES	14th INFANTRY BRIGADE (9.41-10.43) 2Bn Black Watch, 2Bn York & Lancaster Regiment, 1Bn Bedfordshire & Hertfordshire Regiment (9.41-10.43) 16th INFANTRY BRIGADE (9.41-2.42)(3.43-10.43) 2Bn Leicestershire Regiment, 2Bn Queens Royal Regiment, 2Bn Kings Own Royal Regiment (9.41-2.42)(3.43-10.43) 23rd INFANTRY BRIGADE (9.41-10.43) 4Bn Border Regiment, 1Bn Buffs, 1Bn Durham Light Infantry (9.41-1.42), 1Bn South Staffordshire Regiment (2.42-10.43)
SUPPORT	No units

76th INFANTRY DIVISION - Formed by redesignation of Norfolk County Division Sept.1941, disbanded in UK Aug.1944

INSIGNIA Norfolk wherry in full sail in red on black background

THEATRES UK (9.41-8.44)

DIVISIONAL TROOPS 76Indpdt.Coy.-Sqdn.Recce.Corps (1.42-1.43), 52Fld.RA (11.41-8.42), 53Fld.RA (11.41-6.42), 167Fld.RA (5.42-1.43), 173Fld.RA (11.43-8.44), 56Atk.RA (11.41-12.41), 100Atk.RA (1.42-11.42), 94Atk.RA (1.43-11.43), 56Med.RA (1.43-8.44), 60Hvy.RA (7.43-11.43)

BRIGADES 213th INFANTRY BRIGADE (HOME) (9.41-8.44)

13Bn Royal Warwickshire Regiment (9.41-8.42), 9Bn Royal Berkshire Regiment, 14Bn South Staffordshire Regiment (9.41-12.42), 11Bn Worcestershire Regiment (9.42-12.42), 9Bn Kings Own Scottish Borderers, 11Bn York & Lancaster Regiment, 9Bn Seaforth Highlanders (1.43-11.43), 1Bn Foresters (12.43-6.44), 4Bn East Lancashire Regiment (12.43-6.44), 7Bn Border Regiment (12.43-3.44), 5Bn Border Regiment (4.44-6.44), 6Bn East Sussex Regiment, 7Bn Gloucestershire Regiment, 4Bn Oxfordshire & Buckinghamshire Light Infantry (6.44-8.44)

220th INFANTRY BRIGADE (HOME) (9.41-8.44)

9Bn Bedfordshire & Hertfordshire Regiment (9.41-7.42), 9Bn Royal Norfolk Regiment (9.41-12.42), 7Bn Royal Norfolk Regiment (9.41-7.42), 1/5Bn West Yorkshire Regiment (8.42-12.42), 6Bn Northamptonshire Regiment (8.42-9.43), 18Bn Welch Regiment (12.42-9.43), 6Bn Kings Own Yorkshire Light Infantry (12.42-9.43), 5Bn West Yorkshire Regiment (12.42-8.44), 9Bn Royal Welsh Fusiliers, 2/6Bn Lancashire Fusiliers, 6Bn Kings Own Royal Regiment (10.43-8.44)

222nd INFANTRY BRIGADE (HOME) (9.41-11.43)

8Bn Lincolnshire Regiment (9.41-12.42), 11Bn Royal Scots Fusiliers (9.41-9.42), 8Bn Leicestershire Regiment (9.41-5.42), 1Bn Leicestershire Regiment (6.42-12.42), 4Bn East Lancashire Regiment (10.42-12.42), 7Bn East Lancashire Regiment, 7Bn Border Regiment, 1Bn Foresters (1.43-11.43)

SUPPORT No units

77th INFANTRY DIVISION - Formed Nov.1941 by redesignation of Devon & Cornwall County Division. Disbanded in UK Aug.1944

INSIGNIA — Red sword held aloft by a white arm emerging from three blue wavy lines. Black background

THEATRES — UK (11.41-8.44)

DIVISIONAL TROOPS — 77Indpdt.Coy.-Sqdn. Recce.Corps (1.42-1.43), 10Bn Kings Royal Rifle Corps (Motor Bn.) (1.43-11.43), 23Fld.RA (12.41-12.42), 175Fld.RA (7.42-12.42)(8.43-8.44), 176Fld.RA (12.42-8.44), 87Atk.RA (1.42-10.42)

BRIGADES

203rd INFANTRY BRIGADE (HOME) (11.41-8.44)

12Bn Devonshire Regiment, 10Bn Loyal Regiment (11.41-5.42), 9Bn Essex Regiment (11.41-9.42), 2Bn Loyal Regiment (6.42-9.42), 2Bn East Surrey Regiment, 2Bn Royal Sussex Regiment (10.42-12.42), 15Bn Queens Royal Regiment (10.42-10.43), 8Bn Devonshire Regiment, 10Bn Royal Sussex Regiment (1.43-10.43), 11Bn South Staffordshire Regiment (11.43-8.44), 11Bn York & Lancaster Regiment, 9Bn Seaforth Highlanders (11.43-4.44), 11Bn Argyll & Sutherland Highlanders, 2/6Bn Lancashire Fusiliers (5.44-8.44)

209th INFANTRY BRIGADE (HOME) (11.41-8.44)

7Bn & 8Bn Buffs (11.41-9.42), 9Bn & 10Bn Buffs (11.41-10.43), 11Bn Hampshire Regiment, 14Bn Durham Light Infantry (10.42-8.44), 18Bn Welch Regiment, 6Bn Northamptonshire Regiment (10.43-8.44)

211th INFANTRY BRIGADE (HOME) (11.41-1.43)

11Bn Devonshire Regiment, 14Bn Royal Warwickshire Regiment (11.41-8.42), 10Bn East Surrey Regiment (11.41-1.43), 13Bn Queens Royal Regiment (9.42-3.44), 2/6Bn Lancashire Fusiliers (9.42-8.43), 7Bn Royal West Kent Regiment (2.43-3.44), 8Bn Devonshire Regiment (9.43-3.44), 5Bn Kings Shropshire Light infantry, 8Bn Worcestershire Regiment, 8Bn Royal Warwickshire Regiment, 9Bn Royal Welsh Fusiliers (4.44-8.44)

11th TANK BRIGADE (2.43-10.43)

107RAC, 110RAC, 111RAC (2.43-10.43)

SUPPORT — No units

78th INFANTRY DIVISION - Formed in the UK May 1942, still operating Sept.1945

INSIGNIA	Yellow crusaders battle axe on black disc
THEATRES	UK (5.42-10.42), N.Africa (11.42-7.43), Sicily (7.43-9.43), Italy (9.43-7.44)(9.44-5.45), Austria (5.45-8.45)
DIVISIONAL TROOPS	29Indpdt.Sqdn. Recce.Corps (6.42-8.42), 56Regt. Recce.Corps (9.42-12.43), 56th Recce.Regt. RAC (1.44-8.45), 17Fld.RA (6.42-7.45), 132Fld.RA, 138Fld.RA (7.42-8.45), 154Fld.RA (7.45-8.45), 64Atk.RA (7.42-8.45), 49LAA.RA (7.42-11.44)

BRIGADES

1st INFANTRY BRIGADE (GUARDS) (5.42-3.43)

3Bn Grenadier Guards, 2Bn Coldstream Guards (5.42-3.43), 2Bn Hampshire Regiment (5.42-11.42), 8Bn Argyll & Sutherland Highlanders (12.42-3.43), 3Bn Welsh Guards (3.43)

11th INFANTRY BRIGADE (5.42-8.45)

2Bn Lancashire Fusiliers, 1Bn East Surrey Regiment, 5Bn Northamptonshire Regiment (5.42-8.45)

36th INFANTRY BRIGADE (5.42-8.45)

6Bn Royal West Kent Regiment, 5Bn Buffs (5.42-8.45), 7Bn Royal West Kent Regiment((5.42-8.42), 8Bn Argyll & Sutherland Highlanders (9.42-8.45)

38th (IRISH) INFANTRY BRIGADE (4.43-8.45)

2Bn London Irish Rifles, 1Bn Royal Irish Fusiliers (4.43-8.45), 6Bn Royal Inniskilling Fusiliers (4.43-8.44), 2Bn Royal Inniskilling Fusiliers (9.44-8.45)

SUPPORT

MACHINE GUN UNIT (7.44-8.45)

1Bn Kensington Regiment (7.44-8.45)

SUPPORT BATTALION (6.43-7.44)

1Bn Kensington Regiment (6.43-7.44)

<u>80th INFANTRY (RESERVE) DIVISION</u> - Formed in UK Jan.1943, as a training and draftfinding unit. Disbanded Aug.1944

INSIGNIA — Red ocean liner, sea and funnel smoke light blue, yellow background, all within a light blue border

THEATRES — UK (1.43-8.44)

DIVISIONAL TROOPS — 80Regt.Recce.Corps (1.43-12.43), 80th Recce.Regt. RAC (1.44-8.44), 195Fld.RA (1.43-8.44), 199Fld.RA (8.43-10.43), 94Atk.RA (11.43-8.44)

BRIGADES — 45th INFANTRY BRIGADE (1.43-6.44)

10Bn Cameronians, 10Bn Black Watch, 11Bn Argyll & Sutherland Highlanders (1.43-6.44)

211th INFANTRY BRIGADE (1.43-8.44)

10Bn East Surrey Regiment (1.43), 13Bn Queens Royal Regiment (1.43-3.44), 2/6Bn Lancashire Fusiliers (1.43-8.43), 7Bn Royal West Kent Regiment (2.43-3.44), 8Bn Devonshire Regiment (9.43-3.44), 5Bn Kings Shropshire Light Infantry, 8Bn Worcestershire Regiment, 8Bn Royal Warwickshire Regiment, 9Bn Royal Welsh Fusiliers (4.44-8.44)

SUPPORT — MACHINE GUN UNIT (1.43-8.44)

5Bn Cheshire Regiment (1.43-8.44)

1st AIRBORNE DIVISION - Formed in UK November 1941, still operating Sept.1945

INSIGNIA — Bellerophon astride Pegasus in pale blue, maroon background

THEATRES — UK (11.41-4.43), N.Africa (4.43-7.43), Sicily (7.43), N.Africa (7.43-9.43), Italy (9.43-11.43), UK (12.43-9.44), N.W.Europe (9.44), UK (9.44-5.45), Norway (5.45-8.45), UK (8.45)

DIVISIONAL TROOPS — 1st Airlanding Sqdn. Recce. Corps (12.42-12.43), Airborne Light Tank Sqdn. RAC (8.42-4.43), 1st Airlanding Recce. Sqdn. (1.44-8.45) 1st Airlanding Lt.Bty. RA (9.42-2.43), 1st Airlanding Lt. Regt. RA (2.43-10.43)(2.44-8.45), 204Indpdt.Atk.Bty. RA (8.42-10.42), 1st Airlanding Atk.Bty. RA (9.42-3.45), 2nd Airlanding Atk.Bty. RA (10.42-3.45), 5th Airlanding Atk.Bty. RA (2.45-3.45), 1st Airlanding Atk.Regt. RA (3.45-8.45), 283LAA.Bty. RA (7.42-12.42 1st Airlanding LAA.Bty. RA (12.42-2.44)

BRIGADES — 1st AIRLANDING BRIGADE (11.41-8.45)

1Bn Border Regiment, 2Bn South Staffordshire Regiment (11.41-8.45), 2Bn Oxfordshire & Buckinghamshire Light Infantry, 1Bn Royal Ulster Rifles (11.41-5.43), 7Bn Kings Own Scottish Borderers (12.43-8.45)

*Units forming Brigade Group 11.41-3.43

458Lt.Bty.RA, 223Atk.Bty.RA (12.41-6.42), 1st Airlanding Lt.Bty. 1st Airlanding Atk.Bty.RA (7.42-9.42), 1st Airlanding Recce. Coy.-Sqdn. (12.41-12.42)

1st PARACHUTE BRIGADE (11.41-8.45)

1Bn, 2Bn & 3Bn Parachute Regiment (11.41-8.45), 4Bn Parachute Regiment (1.42-7.42)

2nd PARACHUTE BRIGADE (7.42-11.43)

4Bn, 5Bn & 6Bn Parachute Regiment (7.42-11.43)

4th PARACHUTE BRIGADE (6.43-11.44)

10Bn, 156Bn Parachute Regiment (6.43- 11.44), 11Bn Parachute Regiment (1.44-9.44)

SUPPORT — ARMY AIR CORPS - 21 Indpdt. Para. Coy. (7.42-8.45)

6th AIRBORNE DIVISION - Formed in UK May 1943, still operating Sept.1945

INSIGNIA	Bellerophon astride Pegasus in pale blue, maroon background
THEATRES	UK (5.43-9.44), N.W.Europe (9.44), UK (9.44-5.45), Norway (5.45-8.45), UK (8.45)
DIVISIONAL TROOPS	1st Airborne Light Tank Sqdn. (5.43-1.44), 6th Airborne Division Armd. Recce. Regt. (1.44-8.45), 3rd & 4th Airlanding Atk. Btys. (7.43-2.45), 53Atk.RA (10.43-11.43), 2nd Airlanding Atk. Regt. (2.45-8.45), 53Lt.RA (11.43-8.45). 2nd Airlanding LAA Bty. (5.43-2.44)
BRIGADES	6th AIRLANDING BRIGADE (6.43-8.45) 2Bn Oxfordshire & Buckinghamshire Light Infantry, 1Bn Royal Ulster Rifles (6.43-8.45), 12Bn Devonshire Regiment (8.43-8.45) 3rd PARACHUTE BRIGADE (6.43-8.45) 7Bn Parachute Regiment (6.43-9.43), 8Bn & 9Bn Parachute Regiment (6.43-8.45) 5th PARACHUTE BRIGADE (6.43-6.45) 12Bn & 13Bn Parachute Regiment (6.43-6.45), 7Bn Parachute Regiment (9.43-6.45) 2nd PARACHUTE BRIGADE (7.45-8.45) 4Bn, 5Bn & 6Bn Parachute Regiment (7.45-8.45)
SUPPORT	ARMY AIR CORPS - 22Indpdt. Para. Coy. (10.43-7.45)

GUARDS DIVISION - Formed by conversion and redesignation of Guards Armoured Divison, N.W.Europe June to August 1945

INSIGNIA Blue shield edged red with stylised white eye in centre

THEATRES N.W.Europe (6.45-8.45)

DIVISIONAL TROOPS 2nd Household Cavalry Regiment (6.45-8.45), 55Fld.RA, 92Fld.RA, 153Fld.RA, 75Atk.RA, 94LAA.RA (6.45-8.45)

BRIGADES 5th GUARDS BRIGADE (6.45-8.45)

1Bn Grenadier Guards, 2Bn Grenadier Guards, 1Bn Coldstream Guards, 2Bn Irish Guards (6.45-8.45)

6th GUARDS BRIGADE (6.45-8.45)

4Bn Grenadier Guards, 4Bn Coldstream Guards, 3Bn Scots Guards (6.45-8.45)

32nd GUARDS INFANTRY BRIGADE (6.45-8.45)

5Bn Coldstream Guards, 3Bn Irish Guards, 2Bn Scots Guards, 2Bn Welsh Guards

SUPPORT MACHINE GUN UNIT (6.45-8.45)

1st Independent Machine Gun Company (6.45-8.45)

TABLES I to III

Arm of Service Markings and

Vehicle colour schemes

SUPPLEMENTS 1, 2 & 3

Orders of Precedence

SYNOPSIS OF ARM OF SERVICE MARKINGS - EUROPE

TABLE I

Line	Unit	1939-40 (Col 1)		1941-42 (Col 2)		1942-45 (Col 3)		
a.	Divisional Cavalry Regiment	2	Black	-		-		
b.	Armoured Car Regiment	129	Black	47	Black	44	Blue/Green	
c.	Reconnaissance Regiment	-		41	Green/Blue	45	Green/Blue	
d.	Senior Armoured Brigade HQ	3	Red	50	Red	50	Red	
e.	3 Armoured Regiments	4,5,6	Red	51,52,53	Red	51,52,53	Red	
f.	Junior Armoured Brigade HQ	7	Green	60	Green	60	Green	Inf. Bde.
g.	3 Armoured Regiments	8,9,10	Green	61,62,63	Green	61,62,63	Green	
h.	Senior Infantry Brigade HQ	?		81	Red	81	Red	
	3 Infantry Battalions	14,15,16	Red	55,56,57	Red	55,56,57	Red	
i.	Second Infantry Brigade HQ	?		87	Green	87	Green	
j.	3 Infantry Battalions	17,18,19	Green	60,61,62	Green	60,61,62	Green	
k.	Junior Infantry Brigade HQ	?		94	Brown	94	Brown	
l.	3 Infantry Battalions	20,21,22	Brown	67,68,69	Brown	67,68,69	Brown	
m.	Armoured Divn Motor Battalions	13,14	Red	54	Red	54	Red	
n.	Machine Gun Battalion	-		53	Red	64	Black	
o.	Field Regt - Inf Divn (RA)	10,11,12	Red/Blue	42,43,44	Red/Blue	42,43,44	Red/Blue	
p.	Field Regt - Armd Divn (RA/RHA)	11	Red/Blue	74,76	Red/Blue	74,76	Red/Blue	
q.	Anti-Tank Regt - Inf Divn (RA)	13	Red/Blue	46	Red/Blue	46	Red/Blue	
r.	Anti-Tank Regt - Armd Divn (RA)	-		77	Red/Blue	77	Red/Blue	
s.	Anti-Tank Regt - A'Bne Divn (RA)	-		-		47	Red/Blue	
t.	LAA/Anti-Tank Regt - Armd Divn (RA)	12	Red/Blue	-		-		
u.	LAA Regt - Armd Divn (RA)	-		73	Red/Blue	73	Red/Blue	
v.	LAA Regt - Inf Divn (RA)	-		47	Red/Blue	47	Red/Blue	
w.	Light Regt - A'Bne Divn (RA)	-		-		46	Red/Blue	
x.	Armd Recce Regt - A'Bne Divn	-		-		85	Maroon	

SYNOPSIS OF ARM OF SERVICE MARKINGS - MIDDLE EAST — TABLE II

Line	Unit	1939-40 (Col 1)		1941-42 (Col 2)		1942-44 (Col 3)		
a.	Divisional Cavalry Regiment	-		81	Green/Blue	81	Green/Blue	
b.	Armoured Car Regiment	14	White/Green	76	Green/Blue	76	Green/Blue	
c.	Reconnaissance Regiment	-		-		41	Green/Blue	
d.	Senior Armoured Brigade HQ	23	Red	71	Red	71	Red	
e.	3 Armoured Regiments	24,25,26	Red	40,86,67	Red	40,86,67	Red	
f.	Junior Armoured Brigade HQ	27	Green	49	Green	71	Red	up to March 1942
g.	3 Armoured Regiments	28,29,30	Green	65,68,63	Green	40,86,67	Red	up to March 1942
h.	Senior Infantry Brigade HQ	?		65	Red	65	Red	
i.	3 Infantry Battalions	?		91,80,52	Red	91,80,52	Red	
j.	Second Infantry Brigade HQ	?		72	Green	72	Green	
k.	3 Infantry Battalions	?		84,75,92	Green	84,75,92	Green	
l.	Junior Infantry Brigade HQ	?		58	Brown	58	Brown	
m.	3 Infantry Battalions	?		60,62,89	Brown	60,62,89	Brown	
n.	Armd Division Motor Battalions	4	Red	60 Red, 87 Green		60	Red	
o.	Machine Gun Battalion	-		51	Red	51	Red	
p.	Field Regt - Inf Divn (RA)	?		88,71,53	Red/Blue	88,71,53	Red/Blue	
q.	Field Regt - Armd Divn (RA/RHA)	?		77	Red/Blue	77,55	Red/Blue	
r.	Anti-Tank Regt - Inf Divn (RA)	?		96	Red/Blue	96	Red/Blue	
s.	Anti-Tank Regt - Armd Divn (RA)	?		55	Red/Blue	64	Red/Blue	
t.	LAA/Anti-Tank Regt - Armd Divn (RA)	?		-		-		
u.	LAA Regt - Inf Divn (RA)	-		55	Red/Blue	55	Red/Blue	
v.	LAA Regt - Armd Divn (RA)	-		54	Red/Blue	54	Red/Blue	
w.	Infantry Battalions - Armd Divn	-		-		65,68,63	Green	

VEHICLE CAMOUFLAGE COLOURS 1939-45

TABLE III

THEATRE	INTRODUCED	DESCRIPTION	SPECIFICATION
Europe	Early 1939	Middle Bronze Green	BSI 381: 1930 colour 23 (G3)
Europe	Early 1939	Deep Bronze Green	BSI 381: 1930 colour 24 (G4)
Europe	Early 1939	Light Bronze Green	BSI 381: 1930 colour 22 (G5)
Europe	Late 1941	Khaki Brown	BSI 987C: 1942 standard camouflage colour No.2
Europe	Late 1941	Dark Brown	BSI 987C: 1942 standard camouflage colour No.1A
Europe	Late 1941	Light Khaki Brown	BSI 987C: 1942 shade No.4
Europe	Early 1944	Olive Drab	BSI 987C: 1942 colour No.15 (Replaced camfl. colour No.2)
North Africa	Late 1939	Light Stone Pale Cream Deep Cream	BSI 381: 1930 colour No.61 BSI 381: 1930 colour No.52 BSI 381: 1930 colour No.53
North Africa	Early 1940	Sand Light Grey Slate Grey	BSI 381: 1930 colour No.64 BSI 381: 1930 colour No.28 BSI 381: 1930 colour No.34
Tunisia	Mid 1943	Middle Bronze Green	BSI 381: 1930 colour No23 (G3)
Syria	Mid 1943	Sand Terra Cotta (?)	BSI 381: 1930 colour No.64 BSI 987C: 1942 shade No.11A
Far East	Early 1942	Dark Olive Drab	BSI 987C: 1942 shade No.13
Far East	Early 1945	Deep Olive Drab	BSI 987C: 1942 shade No.16

Application of the above colours varied, although the most common variation was a lighter shade with a darker disruptive pattern overpainted. In many instances, vehicles were left in one colour without disruptive patterns, although local mud or clay was used to good effect. A complicated converging stripe pattern was used in N.Africa c.1940 (Caunter camouflage). Terra Cotta was unique to Syria, matching the native red sandstone rock.

REGIMENTAL SENIORITIES AND ORDERS OF PRECEDENCE

(With particular emphasis on their relationship to military vehicle arm-of-service & tactical marking)

SUPPLEMENT ONE

Extract from the Army List, January 1945, page 13-xiii, (Order of Precedence) relative to units detailed in this publication :-

1	Life Guards & Royal Horse Guards
2	Royal Horse Artillery
3	Regiments of Cavalry of the Line
4	Regiments of the Royal Armoured Corps
5	Royal Regiment of Artillery
8	Regiments of Foot Guards
9	Regiments of Infantry of the Line
22	Supplementary Reserve
23	Militia
25	Territorial Army

Note (a) Royal Marines, when serving under the Army Act, take precedence immediately following the Royal Berkshire Regiment.

(b) Units 3,4,8,9 take precedence amongst themselves per supplements Two and Three of this publication. Units 22,23 & 25 take precedence in order of arms as for the Regular Army.

Note The distinction between Regular & Territorial units appears to have been reflected in the markings of RAC vehicles.

This distinction does not appear to have been so well defined in Infantry Regiments where the first two Battalions (three for Foot Guards) could generally be identified as 'regulars' and other Battalions Territorial. The seniority of the Regiment rather than the Battalion seems to have been employed.*

* Hodges P., British Military Markings 1939-45, page 51 (Almark)

Sanders J., British Guards Armoured Division 1941-45, plate A, (Osprey-Vanguard)

SUPPLEMENT TWO
PART ONE

INFANTRY REGIMENTS IN ORDER OF PRECEDENCE

FOOT GUARDS

Grenadier Guards	(GG)
Coldstream Guards	(CG)
Scots Guards	(SG)
Irish Guards	(IG)
Welsh Guards	(WG)

INFANTRY OF THE LINE

The Royal Scots	(RS)
Queens Royal Regiment	(Qns)
Buffs (Royal East Kent Regiment)	(Buffs)
Kings Own Royal Regiment	(KO)
Royal Northumberland Fusiliers	(NF)
Royal Warwickshire Regiment	(War)
Royal Fusiliers	(RF)
The Kings Regiment	(Kgs)
Royal Norfolk Regiment	(Norf)
Lincolnshire Regiment	(Linc)
Devonshire Regiment	(Dev)
Suffolk Regiment	(Suf)
Somerset Light Infantry	(Somt)
West Yorkshire Regiment	(WYks)
East Yorkshire Regiment	(EYks)
Bedfordshire & Hertfordshire Regiment	(BedH)
Leicestershire Regiment	(Leic)
The Green Howards	(GrnH)
Lancashire Fusiliers	(LF)
Royal Scots Fusiliers	(RSF)
Cheshire Regiment	(Ches)
Royal Welsh Fusiliers	(RWF)
South Wales Borderers	(SWB)
+Brecknock Bn.	(Bcn)
Kings Own Scottish Borderers	(KOSB)
Cameronians (Scottish Rifles)	(Cam)
Royal Inniskilling Fusiliers	(Inskn)
Gloucestershire Regiment	(Glos)
Worcestershire Regiment	(Worc)
East Lancashire Regiment	(ELnc)
East Surrey Regiment	(Sury)
Duke of Cornwalls Light Infantry	(DCLI)
Duke of Wellingtons Regiment	(DWR)
Border Regiment	(Bord)
Royal Sussex Regiment	(Susx)
Hampshire Regiment	(Hamp)
South Staffordshire Regiment	(SStfs)
Dorsetshire Regiment	(Dors)
South Lancashire Regiment	(SLnc)
Welch Regiment	(Wch)
Black Watch (Royal Highlanders)	(BW)
+Tyneside Scottish	(TyS)
Oxfordshire & Buckinghamshire Light Infantry	(OxB)
+Buckinghamshire Bns.	(Buk)
Essex Regiment	(Esx)
Sherwood Foresters	(For)
Loyal Regiment	(Loy)
Northamptonshire Regiment	(Nthp)
Royal Berkshire Regiment	(Berk)
Royal West Kent Regiment	(RWK)
Kings Own Yorkshire Light Infantry	(KOYLI)
Kings Shropshire Light Infantry	(KSLI)
Middlesex Regiment	(Midx)
+Kensington Regiment	(Ken)
Kings Royal Rifle Corps	(KRRC)
+Queen Victorias Rifles	(QVR)
+Queens Westminsters	(QW)
+ The Rangers	(Ran)
Wiltshire Regiment	(Wilt)
Manchester Regiment	(Man)
North Staffordshire Regiment	(NStfs)
York and Lancaster Regiment	(YL)

SUPPLEMENT TWO
PART TWO

INFANTRY REGIMENTS IN ORDER OF PRECEDENCE

York and Lancaster Regiment	
+Hallamshire Bn.	(Halm)
Durham Light Infantry	(DLI)
Highland Light Infantry	(HLI)
+Glasgow Highlanders	(GHld)
Seaforth Highlanders	(Sea)
Gordon Highlanders	(Gdn)
+London Scottish	(LSc)
Cameron Highlanders	(CamH)
+Liverpool Scottish	(LvS)
Royal Ulster Rifles	(RUR)
+London Irish Rifles	(LIR)
Royal Irish Fusiliers	(RIF)
Argyll and Sutherland Highlanders	(ASH)
The Rifle Brigade	(RB)
+London Rifle Brigade	(LRB)
+Tower Hamlets Rifles	(THR)
Parachute Regiment	(Para)

SUPPLEMENTARY RESERVE

Cambridgeshire Regiment	(Camb)
Herefordshire Regiment	(Hrfd)
Hertfordshire Regiment	(Hert)
Monmouthshire Regiment	(Mon)
Lovat Scouts	(Lov)

Note : Units anetated '+' are classed as battalions of that Regiment.

Some Title Changes between 1939 & 1945

YH	to	9KOYLI	6SWB	to	158RAC
10RTR	to	7RTR	10Glos	to	159RAC
144RAC	to	4RTR	9Susx	to	160RAC
12DLI	to	TyS	12Grn	to	161RAC
1QVR	to	7KRRC	161RAC	to	161Recce.
2QVR	to	8KRRC	9RWK	to	162RAC
1Ran	to	9KRRC	13For	to	163RAC
2Ran	to	10KRRC	5KO	to	107RAC
1QW	to	11KRRC	1/5LF	to	108RAC
2QW	to	12KRRC	1/6LF	to	109RAC
1LRB	to	7RB	5Bord	to	110RAC
2LRB	to	8RB	5Man	to	111RAC
1THR	to	9RB	9For	to	112RAC
2THR	to	10RB	2/5WYks	to	113RAC
9Loy	to	148RAC	113RAC	to	14WYks
7KOYLI	to	149RAC	2/6DWR	to	114RAC
10YL	to	150RAC	2/7DWR	to	115RAC
10KO	to	151RAC	9Gdn	to	116RAC
11Kgs	to	152RAC	7Buffs	to	141RAC
8Esx	to	153RAC	7Suf	to	142RAC
9NStfs	to	154RAC	9LF	to	143RAC
15DLI	to	155RAC	8ELnc	to	144RAC
11HLI	to	156RAC	8DWR	to	145RAC
9Hamp	to	157RAC	9DWR	to	146RAC
10Hamp	to	147RAC	8RWK	to	9Med.RA
16RF	to	10Med.RA	9Esx	to	11Med.RA

Note : This list is not considered to be a comprehensive inventory of change

SUPPLEMENT THREE

ARMOURED REGIMENTS IN ORDER OF PRECEDENCE

HORSE GUARDS

Household Cavalry Regiment	(HCR)

REGIMENTS OF HORSE

1st (Kings) Dragoon Guards	(KDG)
The Queens Bays (2nd Dragoon Guards)	(Bays)
3rd Carabineers (Prince of Wales's Dragoon Guards)	(3DG)
4th/7th (Royal) Dragoon Guards	(4/7DG)
5th Royal Inniskilling Dragoon Guards	(5DG)

CAVALRY OF THE LINE

1st The Royal Dragoons	(Royals)
Royal Scots Greys (2nd Dragoons)	(Greys)
3rd The Kings Own Hussars	(3H)
4th The Queens Own Hussars	(4H)
7th The Queens Own Hussars	(7H)
8th Kings Royal Irish Hussars	(8H)
9th Queens Royal Lancers	(9L)
10th Royal Hussars (Prince of Wales' Own)	(10H)
11th Hussars (Prince Alberts Own)	(11H)
12th Royal Lancers (Prince of Wales')	(12L)
13/18th Royal Hussars (Queen Marys' Own)	(13/18H)
14/20th Hussars	(14/20H)
15/19th Kings Royal Hussars	(15/19H)
16/5th Lancers	(16/5L)
17/21st Lancers	(17/21L)
22nd Dragoons	(22D)
23rd Hussars	(23H)
24th Lancers	(24L)
25th Dragoons	(25D)
26th Hussars	(26H)
27th Lancers	(27L)

REGULAR ROYAL TANK REGIMENTS

Nos. 1 to 12 RTR (by number)	(No.* RTR)

CAVALRY MILITIA

Northern Irish Horse	(NIH)

YEOMANRY CAVALRY REGIMENTS

Royal Wiltshire Yeomanry	(RWY)
Warwickshire Yeomanry	(WY)
Yorkshire Hussars	(YH)
Notts (Sherwood Rangers) Yeomanry	(NSR)
Staffordshire Yeomanry	(SY)
Cheshire Yeomanry	(CY)
Yorkshire Dragoons	(YD)
North Somerset Yeomanry	(NSY)
Derbyshire Yeomanry	(DY)
Royal Gloucestershire Hussars	(RGH)
Lothian and Border Horse	(LBH)
Fife and Forfar Yeomanry	(FFY)
Westminster Dragoons	(WD)
3rd County of London Yeomanry	(3CLY)
4th County of London Yeomanry	(4CLY)
Northamptonshire Yeomanry	(NY)
East Riding Yeomanry	(ERY)
Inns of Court Regiment	(IC)

TERRITORIAL ROYAL TANK REGIMENTS

Nos. 40 to 51 RTR (by number)	(No.* RTR)

INFANTRY CONVERTED TO ARMOUR

Guards Regiments (as for Foot Guards)	(per Supp. 2)
RAC Regiments by RAC Number	(No.* RAC)

INDEX

REGIMENTS TO DIVISIONS

IN SEQUENCE :-

Armoured
Infantry
Artillery
Reconnaissance
Assault R.E.

ABREVIATIONS :

Abreviations for Armoured and Infantry Regiments are described in Supplements 1,2 & 3, other abreviations used are described below.

Divisions:

(C) Cavalry, (A) Armoured, (G) Guards
(I) Infantry (LI) London Infantry,
(AB) Airborne, (GA) Guards Armoured

Artillery:

(ATk) Anti-Tank, (LAA) Light Anti-Aircraft,
(LAA.ATk) Combination of above units

INDEX - REGIMENTS TO DIVISIONS - ARMOURED/INFANTRY A-D

HCR	1C,10A,GA G
KDG	2A
Bays	1A
4/7DG	9A,79A,2I
5DG	7A,4I
Royals	1A,10A,
Greys	1C,10A,79A 6I
3H	2A,7A,10A
4H	1A,2A
7H	7A,10A
8H	7A
9L	1A
10H	1A
11H	7A
12L	1A
13/18H	9A,79A,1I
15/19H	11A,3I
16/5L	6A
17/21L	6A
22D	11A,42A, 79A
23H	11A
24L	11A
27L	11A
NIH	23I,1I
RWY	1C,10A
WY	1C,10A
YH	61I,1C, 10A,50I
NSR	1C,10A
SY	1C,10A, 79A
CY	1C
YD	1C,1A
NSY	1C
1DY	6A

2DY	8A,10A,51I
1RGH	9A,6A
2RGH	1A,2A,7A
1LBH	11A,42A, 79A,48I
2LBH	6A
1FFY	79A,51I
2FFY	11A
WD	11A,42A, 79A
3CLY	1A,2A,7A
4CLY	1A,2A,7A
1NY	6A,42A,79A
2NY	6A,11A,79A
ERY	9A,79A
IC	9A
1RTR	7A
2RTR	1A,7A,10A
3RTR	1A,2A,7A, 10A,11A
4RTR	7A
5RTR	1A,2A,7A
6RTR	2A,7A,10A
7RTR	7A,79A,53I
8RTR	7A,10A
9RTR	53I
10RTR	53I
11RTR	79A,43I
12RTR	4I
40RTR	8A,10A
41RTR	8A
42RTR	7A,79A
43RTR	79A,3I
44RTR	7A
45RTR	8A
46RTR	8A,10A
47RTR	8A
48RTR	4I

49RTR	79A
50RTR	8A,10A
51RTR	1I,43I
49APC	79A
107RAC	77I
108RAC	42A
109RAC	42A
110RAC	77I
111RAC	77I
112RAC	42A
141RAC	79A
142RAC	1I,43I
144RAC	3I,79A
145RAC	42A,4I
147RAC	43I,1I
148RAC	79A,3I
151RAC	43I
152RAC	79A
155RAC	79A
162RAC	79A

ASH	
1Bn	6I,8I
2Bn	15I
7Bn	51I
7/10Bn	51I
8Bn	51I,78I
10Bn	9I
11Bn	9I,15I,45I 51I,77I, 80I
Bcn	see SWB
BedH	
1Bn	6I,8I,54I, 70I

BedH	
2Bn	4I,54I
5Bn	18I
6Bn	54I
9Bn	55I,76I
Buffs	
1Bn	1I,56I,6I, 10A,1A,7A, 8I,70I
2Bn	36I
4Bn	2LI,10A
5Bn	78I,12I
7Bn	77I
8Bn	77I
9Bn	77I,55I
10Bn	77I
Bord	
1Bn	1AB,42I
2Bn	36I
4Bn	70I,2LI, 47I,6I
5Bn	76I,38I, 42I
6Bn	66I,54I
7Bn	76I,66I, 54I
Berk	
1Bn	2I
4Bn	3I,48I
5Bn	54I
6Bn	61I
7Bn	54I
9Bn	76I
10Bn	1LI,50I, 56I

Buk	see OxB
BW	
1Bn	4I,51I
2Bn	70I,6I
4Bn	51I,55I, 47I
5Bn	9I,51I
6Bn	4I,51I
7Bn	9I,51I
8Bn	47I
10Bn	15I,48I, 80I
TyS	23I,49I
Cam	
1Bn	2I,18I, 36I
2Bn	5I,18I
6Bn	52I
7Bn	15I,52I
9Bn	15I
10Bn	15I,48I, 80I
Camb	
1Bn	18I
2Bn	18I
CamH	
1Bn	2I
2Bn	6A
4Bn	15I,51I
5Bn	9I,51I
1LvS	55I
2LvS	55I
CG	see GUARDS

Ches	
1Bn	11A
2Bn	61I,50I
4Bn	61I,56I, 1I
5Bn	38I,53I, 80I
6Bn	44I,56I
7Bn	5I
DCLI	
1Bn	47I,55I 61I
2Bn	4I
4Bn	45I
5Bn	43I,45I
Dev	
1Bn	36I
2Bn	7A,50I
4Bn	45I,55I, 61I
6Bn	47I,45I
8Bn	45I,77I, 80I
9Bn	45I
11Bn	77I
12Bn	6AB,77I
DLI	
1Bn	6I,70I,45I
2Bn	2I
6Bn	50I
8Bn	50I
9Bn	7A,50I
10Bn	23I,49I
11Bn	23I,49I
12Bn	23I

INDEX - REGIMENTS TO DIVISIONS - INFANTRY D-K

Regiment	Battalion	Divisions
DLI	14Bn	77I,45I
	16Bn	46I
	18Bn	55I
Dors	1Bn	50I
	2Bn	2I
	4Bn	43I
	5Bn	43I
	9Bn	45I,47I
DWR	1Bn	3I
	6Bn	49I
	1/6Bn	49I
	2/6Bn	46I
	7Bn	49I
	1/7Bn	49I
	2/7Bn	46I
	10Bn	55I,48I
ELnc	1Bn	42I,42A,2I, 53I
	2Bn	36I
	4Bn	76I,42I,38I
	5Bn	66I,59I,42I
	7Bn	76I
Esx	1Bn	36I
	2Bn	2LI,47I,49I
	1/4Bn	54I
	2/4Bn	54I,47I
	5Bn	5I
Esx	1/5Bn	54I
	2/5Bn	54I
	9Bn	77I
EYks	2Bn	3I
	4Bn	50I
	5Bn	23I,50I
	7Bn	54I
For	1Bn	76I,61I,8I
	2Bn	1I
	5Bn	46I
	1/5Bn	18I
	2/5Bn	46I
	8Bn	49I
	9Bn	1A,46I
	14Bn	1A,7A,8A,1I 45I, 10A
	16Bn	54I
Gdn	1Bn	51I,1I
	2Bn	15I
	5Bn	51I
	5/7Bn	51I
	6Bn	51I,1I
	7Bn	9I
	9Bn	9I
	1LSc	56I,1LI,50I
	2LSc	48I,2LI,47I
GHld		see HLI
Glos	2Bn	3I,47I,48I, 49I
	5Bn	48I
	7Bn	76I,61I,47I
	10Bn	36I
GG		see GUARDS
GrnH	1Bn	5I
	4Bn	50I
	5Bn	50I
	6Bn	23I,50I
	7Bn	23I,50I
	10Bn	55I
	11Bn	48I
	12Bn	11A
GUARDS		
GG	1Bn	3I,GA,G
	2Bn	3I,GA,G
	3Bn	78I,6A,1I
	4Bn	15I,GA,G
	5Bn	1I
	6Bn	10A
CG	1Bn	3I,G,GA
	2Bn	78I,6A,1I, 56I
	3Bn	1A,10A,6I
	4Bn	15I,GA,G
	5Bn	GA,G
SG	1Bn	56I,1I
	2Bn	6I,1A,10A, G,GA
	3Bn	15I,GA,G
	4Bn	GA
IG	1Bn	1I,GA
	2Bn	G
	3Bn	GA,G
WG	1Bn	GA
	2Bn	GA,G
	3Bn	78I,6A
Halm		see YL
Hrfd	1Bn	53I,11A
	2Bn	38I
HLI	1Bn	53I,42I,42A
	5Bn	52I
	6Bn	52I
	10Bn	15I
	11Bn	15I
	13Bn	42A
	1GHld	52I
	2GHld	15I
Hamp	2Bn	78I,46I,1I
	1/4Bn	43I,46I
	2/4Bn	4I,43I,46I
	5Bn	43I,46I
	7Bn	43I
Hamp	11Bn	77I,45I
	12Bn	45I
Hert	1Bn	1I
	2Bn	54I
Inskn	2Bn	78I,47I,5I, 48I,38I,55I
	5Bn	48I,38I,55I
	6Bn	78I,6A,1I, 47I
Ken		see Midx
Kgs	2Bn	4I
	5Bn	54I,55I
	8Bn	66I,54I
	9Bn	55I
KO	2Bn	70I,6I
	5Bn	42I
	6Bn	48I,47I, 76I, 54I
	9Bn	47I
KOSB	1Bn	3I
	4Bn	52I
	5Bn	52I
	6Bn	15I
KOSB	7Bn	15I,1AB
	9Bn	76I
KOYLI	1Bn	5I
	1/4Bn	49I
	2/4Bn	46I
	6Bn	76I,47I
	9Bn	1A,7A,1I
KRRC	1Bn	1A,6A,7A, 8A
	2Bn	61I,1A
	7Bn	79A
	8Bn	9A
	9Bn	7A
	10Bn	77I,6A
	11Bn	10A
	12Bn	11A,42A
	1QVR	1LI
	2QVR	9A,2LI
	1Ran	2A
	2Ran	6A
	1QW	8A,1LI
	2QW	11A,2LI
KSLI	1Bn	1I
	2Bn	3I,45I,79A
	4Bn	53I,11A
	5Bn	80I,77I, 38I

INDEX - REGIMENTS TO DIVISIONS - INFANTRY L-R

Regiment	Battalion	Divisions
Leic	1Bn	49I,76I, 54I
	2Bn	6I,8I,36I 70I
	1/5Bn	49I
	2/5Bn	46I
	8Bn	76I
LF	2Bn	78I,4I
	1/5Bn	42I
	2/5Bn	66I,59I
	1/6Bn	42I
	2/6Bn	47I,45I, 59I,66I, 76I,77I,80I
	1/8Bn	2I,42I
	2/8Bn	66I,55I
	9Bn	42I
	11Bn	1I
Linc	2Bn	3I
	4Bn	49I
	6Bn	46I
	8Bn	76I
LIR		see RUR
Loy	1Bn	1I
	2Bn	77I,55I
	5Bn	18I
	6Bn	59I,2I
	10Bn	77I
LRB		see RB
LSc		see Gdn
LvS		see CamH
Man	1Bn	55I,53I
	2Bn	2I,36I
	5Bn	55I,42I
	6Bn	66I,55I
	7Bn	66I,55I, 52I
	8Bn	55I,42I
	9Bn	46I
	1/9Bn	52I
Mon	2Bn	53I
	3Bn	53I,11A
	4Bn	38I
Midx	1Bn	15I
	2Bn	3I
	1/7Bn	51I
	2/7Bn	46I,56I,1I
	8Bn	43I
	1/8Bn	43I
	2/8Bn	61I
	1 Ken	78I,56I
	2 Ken	49I
NF	2Bn	4I,46I
	4Bn	50I
	7Bn	59I
	8Bn	3I,23I
	9Bn	23I
Norf	2Bn	79A,2I,3I
	4Bn	18I
	5Bn	18I,48I
	6Bn	18I
	7Bn	76I,59I
	9Bn	76I,47I
NStfs	2Bn	1I
	6Bn	59I
	7Bn	61I,59I, 47I
Nthp	1Bn	36I
	2Bn	5I
	4Bn	61I
	5Bn	78I,4I
	6Bn	77I,76I, 45I
OxB	1Bn	48I,42I,4I 53I
	2Bn	6AB,1AB
	4Bn	76I,47I, 48I
	5Bn	61I
	6Bn	47I
	7Bn	1LI,56I
	1Buk	54I
	2Buk	61I
Para	1Bn	1AB
	2Bn	1AB
	3Bn	1AB
	4Bn	6AB,1AB
	5Bn	6AB,1AB
	6Bn	6AB,1AB
	7Bn	6AB
	8Bn	6AB
	9Bn	6AB
	10Bn	1AB
	11Bn	1AB
	12Bn	6AB
	13Bn	6AB
	156Bn	1AB
Qns	2Bn	8I,6I,36I 70I
	1/5Bn	7A
	2/5Bn	1LI,12I, 56I
	1/6Bn	7A,50I
	2/6Bn	1LI,12I, 56I
	1/7Bn	7A,50I
	2/7Bn	1LI,12I, 56I
	13Bn	80I,77I, 47I
	15Bn	77I
QVR		see KRRC
QW		see KRRC
Ran		see KRRC
RB	1Bn	1A
	2Bn	6A,7A,8I, 10A
	7Bn	6A,8A
	8Bn	11A
	9Bn	1A
	10Bn	6A
	1LRB	8A
	2LRB	2LI
	1THR	2A
	2THR	6A
	3THR	5I
RF	2Bn	4I
	8Bn	1LI,56I
	9Bn	1LI,56I
	11Bn	2LI,47I
	12Bn	2LI,47I
	15Bn	56I
	16Bn	47I
	17Bn	47I,48I
	18Bn	56I
	19Bn	47I
	20Bn	47I
RIF	1Bn	1I,6A,78I, 45I
	2Bn	55I
	6Bn	47I
RS	1Bn	2I
	2Bn	1I
	7/9Bn	52I
	8Bn	15I
RSF	1Bn	36I
	2Bn	3I
	4/5Bn	52I
	6Bn	15I
	11Bn	76I,49I
RUR	1Bn	6AB,1AB
	2Bn	3I
	7Bn	45I,48I
	8Bn	11A
	1LIR	50I,56I
	2LIR	1I,6A,2LI, 78I,47I
RWF	1Bn	2I
	2Bn	36I
	4Bn	53I
	6Bn	53I
	7Bn	53I
	8Bn	38I

INDEX - REGIMENTS TO DIVISIONS - INFANTRY R-Y/ARTILLERY

ARTILLERY

INDEX - REGIMENTS TO DIVISIONS - ARTILLERY(CONTD.)/RECCE/ASSLT.RE

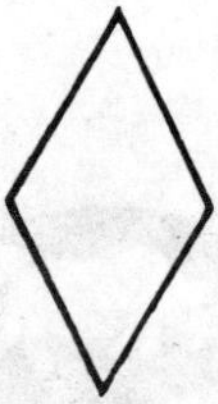

12th (Eastern) Division

51st (Highland) Division

49th (West Riding) Division

42nd (East Lancs) Division
42nd Armoured Division

1st Division

18th Division

78th Division

1st Armoured Division

7th Armoured Division

6th Armoured Division

11th Armoured Division

8th Armoured Division

9th Armoured Division

79th Armoured Division